About Violence

Ethics and Policies, Volume 2

Artur Victoria

Published by Artur Victoria, 2025.

Front cover image by Auhor

Book design by Author

ABOUT VIOLENCE

First edition. April 1, 2025.

ISBN: 979-8230622963

Written by Artur Victoria.

To my wife Karen Candeias, my sons Nicola Alexandra, José Miguel, Filipe Artur, Alexandre João and my grand sons Pedro, Filipe, Tomás, Rodrigo, José Maria. Thanks for your always support and love.

TABLE OF CONTENTS

The ethical dimensions of collective responsibility

Loss of individual responsibility

- Blame shifting
- Inefficiencies in decision-making

The role of education in preventing violence

- The power of social connections
  - Building supportive relationships
  - Addressing social isolation and exclusion
  - Understanding individual violence
  - The role of substance misuse in violent behaviour

Anger management and impulse control

Social Inequality and Community Violence
Economic deprivation and its impact

The role of exclusion and discrimination

Gang dynamics and violent crime

The strategies used by gangs to defend their territories vary widely.

Gang recruitment and socialisation

Societal impacts of violence Erosion of trust and social cohesion

The role of cultural norms and values

Hate crime and bias-motivated violence

Cultural interventions to reduce violence

Individual trauma and its long-term effects

The economic cost of violence

Defining verbal violence and its scope

Enforcement - Delays in investigation and prosecution

The impact of delays on witnesses is of particular concern.

The erosion of public trust in institutions

Improving investigative capacity and effectiveness

Strengthening prosecution and ensuring fair trials

Policy recommendations for reducing violence

Strengthening police investigative capacity is also essential.

Bibliography

Foreword

Foreword Our modern world is one of interconnectedness, where collective action shapes our social, political and economic landscapes. From the work of small teams to the decisions of international bodies, the concept of collective responsibility is central to understanding outcomes. This book offers a comprehensive examination of collective responsibility, giving equal weight to its advantages and disadvantages. We begin by providing a clear definition of collective responsibility, distinguishing it from individual responsibility and exploring its philosophical underpinnings. Subsequent chapters examine both the advantages and the disadvantages. We consider the increased unity, ethical behaviour and fair distribution of workloads that shared responsibility can bring, as well as the risks of diffusion of responsibility, groupthink and blame. It also critically examines the inequitable distribution of consequences, where individuals not directly involved may still face repercussions. The book also assesses how the search for consensus can lead to inefficiencies in decision-making processes. It suggests strategies for striking a balance: promoting clear accountability, ensuring fair distribution of consequences, and encouraging open communication. We argue for a nuanced understanding that recognises the significant potential of collective responsibility, while acknowledging its inherent complexities and potential pitfalls. A supportive and positive leader can foster a collaborative and optimistic environment, while a negative and critical leader can create an atmosphere of stress and negativity, affecting employee morale and performance. This highlights the importance of emotional intelligence in leadership and the responsibility of the leader to cultivate a positive and productive emotional environment.

The impact of emotional contagion extends beyond the immediate group. In the context of social movements, emotional contagion plays a crucial role in galvanising support and mobilising collective action. For example, the collective anger and indignation experienced by participants in a protest can spread rapidly through the crowd, increasing determination and the likelihood of collective action. This is evident in historical examples of social uprisings, where the contagious spread of outrage and shared purpose acted as a powerful catalyst for significant societal change. The visual display of anger, the shared chants and slogans, and the collective experience of oppression all contribute to this intensification of emotion, transforming individual acts of dissent into a cohesive movement.

By navigating this complexity, we can harness the power of collective action while mitigating the risks associated with shared responsibility to create a more just and equitable society.

Violence is a global crisis that transcends geographical boundaries and socio-economic status. Its insidious tendrils reach into every corner of society, leaving a trail of destruction in its wake.

Understanding its complexity is the first step towards building a more peaceful future. This book embarks on a journey into the heart of violence, dissecting its multifaceted nature and exploring the intricate web of factors that contribute to its existence. From the psychological underpinnings of aggression to the social and environmental influences that shape our behaviour, we will delve into the core issues that drive acts of violence. We will explore the critical role of emotional regulation and examine practical techniques that enable individuals to deal constructively with anger, frustration and conflict. This work emphasises the importance of fostering emotional intelligence, both in individuals and in our communities. It also highlights the vital role of education in conflict resolution and the power of strong social support systems in creating resilient communities. Addressing mental health issues, including trauma, through accessible and effective resources is also paramount. The book advocates the implementation of trauma-informed care and the reduction of the stigma associated with mental health. Finally, it examines the importance of legal and social interventions, including stricter gun control and anti-bullying laws, to create a safer environment for all. This book is not just an academic exploration; it's a call to action, urging a unified societal effort to prioritise prevention strategies and invest in resources that address the root causes of violence. By working together - families, schools, communities and governments - we can build a more peaceful and just world for future generations.

Violence in its myriad forms remains a persistent and devastating challenge for individuals, communities and societies worldwide.

To unravel the complexity of violence, it is necessary to distinguish between the individual act and the broader context of aggression at the community level, each of which requires different perspectives and approaches. While recognising the interconnectedness of these forms, we will make a clear distinction between individual violence, rooted in psychological factors and personal experience, and community violence, fuelled by societal inequalities, structural issues and cultural dynamics.

We seek to explore in depth the psychological roots of individual violence, including mental illness, anger management issues and the significant role of substance abuse in triggering or escalating violent behaviour. We analyse specific examples, delving into the motivations behind acts of aggression to illuminate the complex interplay of factors that contribute to such behaviour. We then broaden our scope to examine the societal factors that give rise to community violence, including social inequalities, economic hardship, gang activity and the influence of cultural and ideological differences. The analysis will include a deep dive into the complex dynamics of gang violence and disputes, highlighting the impact of community disorganisation. The discussion will also extend to the profound consequences of violence at both individual and societal levels, examining the long-term physical and psychological effects on victims, the economic costs, and the erosion of trust and social cohesion within communities. We will outline comprehensive strategies for violence prevention and intervention, emphasising the need for multi-stakeholder collaboration and highlighting the effectiveness of community-based initiatives, educational programmes and policy recommendations. Ultimately, we hope to contribute to a more informed and effective approach to reducing violence and building safer, more peaceful communities for all.

Abstract thinking and perspectives

Imagination is a fundamental aspect of human communication, making it easier to understand and consider the points of view of others. It is the ability to put yourself in someone else's shoes, taking into account their thoughts, feelings and experiences. This skill is known as cognitive flexibility, and it is crucial for fostering empathy and effective communication. Cognitive flexibility also plays an important role in the decision-making process. It allows individuals to consider different perspectives, weigh different options and anticipate possible outcomes. This ability is particularly important in today's complex and fast-changing world, where decisions need to be made quickly and compliance is key. Cognitive flexibility enables individuals to make more informed and balanced decisions that take into account multiple perspectives. Furthermore, cognitive flexibility is not only essential for decision-making, but also contributes to personal growth and development. By being open-minded and considering different perspectives, individuals can gain a better understanding of the world around them and expand their own beliefs and values. This can lead to improved relationships, increased empathy and enhanced communication skills. Overall, perspective taking and cognitive flexibility are critical skills that enable individuals to navigate the complexities of life and form meaningful connections with others.

Imagination and cognitive flexibility are not only beneficial in personal relationships, but also in professional settings. By broadening our understanding of the world and challenging our own beliefs and values, we are able to approach problems and situations from different angles. This allows us to generate new ideas, evaluate information more critically and come up with creative solutions. In an ever-changing and evolving world, these skills are essential for success and growth. Engaging in perspective taking and cognitive flexibility can also lead to an increased sense of empathy towards others. By putting ourselves in someone else's shoes, we are able to better understand their perspective and experiences.

This not only improves our relationships with others, but also allows us to communicate more effectively. We are able to consider different points of view and communicate our own thoughts and ideas more clearly. This can be particularly important in diverse and multicultural environments where people may have different beliefs and backgrounds. By broadening our perspectives, we can bridge the gap and make meaningful connections with others.

The ability to cultivate abstract thinking by broadening our perspectives not only allows us to think more deeply about our own thoughts and beliefs, but also helps us to understand and empathise with others. By exploring alternative points of view, we can challenge and expand our own thinking, leading to greater self-awareness and growth.

This is particularly important in diverse and multicultural environments, where navigating and understanding different perspectives is essential to making meaningful connections with others. By broadening our perspectives, we can bridge the gap and form strong bonds with those around us.

Abstract thinking also plays a crucial role in the development of metacognition, which involves reflecting on one's own thought processes. By broadening our perspective, we are able to think about abstract concepts and assess the validity of our own thoughts, becoming more aware of our own biases and assumptions and challenging them in order to form more rounded and informed opinions. In addition, abstract thinking allows us to think critically and creatively, which can be essential for problem solving and decision making. By broadening our perspectives, we are able to approach challenges and conflicts with a more open and flexible mindset, leading to more effective and innovative solutions.

Abstract thinking plays a central role in developing our intellectual skills and broadening our world view. It enables us to analyse complex problems and make connections between seemingly unrelated concepts. By thinking abstractly, we are able to break down problems into smaller, more manageable pieces, allowing us to gain a deeper understanding of the underlying principles and relationships involved.

Abstract thinking is also essential for developing our creativity and critical thinking skills. It allows us to look at problems from different perspectives and come up with innovative solutions. By broadening our perspectives, we are able to explore new ideas and think outside the box, leading to breakthroughs and advances in various fields. This is essential in problem solving and decision making as it allows us to consider all possible options and make informed choices to develop well rounded and informed opinions.

It enables us to think critically and creatively, leading to effective problem solving and decision making. By broadening our perspectives and approaching challenges with an open and flexible mindset, we are able to find innovative solutions and make meaningful contributions in various fields.

Philosophy allows us to delve deeper into complex concepts and ideas through abstract thinking. This type of thinking enables us to go beyond superficial understanding and analyse the underlying principles that govern our world. By examining and questioning fundamental truths, we gain a broader understanding of the world and ourselves.

Unlike concrete thinking, which focuses on the here and now, abstract thinking involves thinking about possibilities and possible outcomes. This type of thinking is the key to developing well-rounded and informed opinions. It allows us to consider multiple perspectives and the implications of different actions. By extending our thinking beyond the immediate circumstances, we can make more effective decisions and solve problems creatively.

By considering different approaches and solutions, we are able to broaden our perspectives and find innovative solutions in different areas of life. This enables us to think critically and creatively and to make meaningful contributions to our field. In addition, abstract thinking allows us to think beyond the obvious and see things from different angles, helping to question and challenge existing ideas, leading to new and improved solutions. By constantly engaging in abstract thinking, we are able to adapt and evolve in an ever-changing world, making us more resilient and open-minded individuals, able to broaden our horizons and make meaningful contributions to society.

Through our ability to think abstractly, we are able to make connections and generalise information, creating abstract categories that encompass a wide range of examples. This helps us to organise and make sense of the vast amount of knowledge that exists in the world. By understanding the underlying concepts and principles, we are able to apply them to different situations and make informed decisions.

By looking at the bigger picture and considering all the possibilities, we are able to push the boundaries of what is possible and make progress in different aspects of our lives. This not only benefits us individually, but also contributes to the growth and development of society as a whole.

Abstract thinking can be summarised as a powerful skill that not only helps us to understand the world around us, but also allows us to shape it by broadening our horizons and thinking beyond boundaries, making meaningful contributions to society and driving progress and growth. Our ability to generalise and abstract information is fundamental to organising knowledge and finding innovative solutions. It is a valuable tool that enables us to maximise our potential and make a positive impact on the world.

A perfect example of abstract thinking can be seen in various forms of symbolic artistic expression. In the arts, it is seen in paintings, sculptures and other forms of creative expression that rely on non-representational forms to convey emotions, ideas or concepts. By moving away from traditional and realistic representations, artists are able to convey deeper and more complex messages through their work, challenging viewers to think beyond the obvious and use their own abstract thinking skills.

Abstract thinking goes beyond traditional methods of problem solving and requires a unique and creative approach. By embracing it, individuals are able to approach challenges with a fresh perspective, leading to innovative and effective solutions. This type of thinking encourages the exploration of different possibilities and is central to tackling societal issues. It encourages individuals to question the norm and challenge conventional beliefs, paving the way for progress and growth. It not only helps solve current problems, but also prevents future ones. It plays a vital role in pushing boundaries and expanding our minds, leading to new discoveries, advances and breakthroughs. Through continuous stimulation, we can make a lasting impact on the world and shape a better future for generations to come.

As we develop this ability, we are able to make connections between seemingly unrelated concepts, leading to new discoveries and advances. By challenging traditional ways of thinking, it is possible to reach new heights and make lasting contributions to the world. This is especially important in today's rapidly changing world, where new challenges and problems arise every day.

Encouraging and developing this ability can create a society that values creativity, innovation and critical thinking. This will lead not only to personal growth and fulfilment, but also to a better and more sustainable future for all - the possibilities are endless.

Sharing ideas - The circulation of cultural narratives and social interactions are powerful catalysts for shaping our collective imagination. When we engage in dialogue and share ideas within a community, we are exposed to diverse perspectives that broaden our creative horizons. This process of sharing creates a social environment that fosters creativity and innovation, allowing new possibilities to be imagined and a shared vision for the future to be envisioned.

Interactions not only broaden our understanding of the world, but also contribute to the development of cultural narratives that define us as a community. Through storytelling and discourse, we construct a shared narrative that reflects our values and beliefs. These cultural narratives not only foster a sense of belonging, but also provide a framework for interpreting the world around us. By engaging in dialogue and building on each other's ideas, we contribute to the evolution of these narratives and enrich our collective imagination.

The contribution of ideas within a community also helps to forge a collective identity; we come together to shape our shared vision for the future. By engaging in meaningful conversation and debate, we don't just shape the imagination, we shape the culture and society in which we live.

In the process, we constantly evolve and expand our understanding of the world and our own identities. This sharing of ideas and perspectives not only stimulates creativity and innovation, but also fosters a sense of connectedness and belonging within a community. By harnessing the power of cultural diffusion and social interaction, we can continue to shape a more vibrant and imaginative future for ourselves and our societies.

By listening to and learning from each other's unique perspectives, we gain a deeper understanding of the world around us. This diversity of thought and experience broadens our collective imagination and enriches the cultural narratives that define us. Through this process of exchange, we also develop empathy and respect for different points of view, creating a more harmonious and accepting community. Moreover, the act of sharing and exchanging ideas within our community not only broadens our own understanding, but also contributes to the growth and progress of society as a whole. By challenging and questioning existing beliefs and norms, we open ourselves up to new possibilities and pave the way for innovation and change. By continually engaging in these discussions and debates, we shape the cultural landscape and push the boundaries of what is possible. In this way, our community becomes a hub of creativity and progress, driven by the diverse perspectives and ideas of its members.

Within a society, the exchange of ideas through discussion and debate is the path to innovation and change. This ongoing engagement shapes the cultural landscape and pushes the boundaries of what is possible. As a result, the community becomes a hub of creativity and progress, fuelled by the diverse perspectives and ideas of its members. Collaborative problem solving is a key benefit of sharing ideas within a community.

When people come together to share their thoughts and experiences, they can pool their collective knowledge and creativity to address challenges and find innovative solutions. This not only benefits the community, but also has a ripple effect throughout society. In addition to solving problems, the exchange of ideas also contributes to cultural enrichment. Ideas are often shaped by cultural influences, and when individuals share their traditions, beliefs and practices, it fosters a rich and diverse collective imagination. This cultural exchange not only broadens our understanding and appreciation of other cultures, but also leads to the creation of new and unique ideas. It also enriches our cultural frameworks.

The ongoing exchange of ideas within a society fosters a sense of mutual inspiration. As individuals share their unique perspectives and thoughts, they not only broaden their understanding of other cultures, but also inspire others to think creatively and push the boundaries of imagination. This collective inspiration fuels a cycle of idea generation and innovation, leading to progress and growth within the community. The exchange of ideas also plays a vital role in the formation of shared values within a society. When people from different backgrounds come together to share ideas, they also share their beliefs and values. Through discussion and debate, a shared set of values can emerge, shaping the collective identity and vision of the community. This not only promotes unity and understanding, but also helps to build a strong and cohesive community.

The dissemination of cultural narratives and social identities is essential to the development of a community's history, present and future. These ideas and stories form a collective narrative that unites community members and shapes their shared vision. When communities foster a culture of sharing ideas, they become catalysts for innovation and creativity. The exchange of ideas allows for the cross-pollination of thoughts, leading to the birth of new concepts, projects and initiatives. This environment of collaboration and cooperation benefits the entire community, leading to progress and growth.

As these narratives and ideas continue to spread, they shape the identity of the community and create a sense of belonging among its members. This shared narrative becomes a source of inspiration and guidance, directing the community's collective imagination towards a common goal. In this way, narrative building and the exchange of ideas play a vital role in the development and evolution of a successful society.

In today's world, civic engagement is an essential part of developing a strong and vibrant community. When individuals come together to share their ideas and work towards a common goal, it fosters a sense of belonging and collective responsibility. Through open and honest discussion, community members can gain a better understanding of the issues they face and work towards solutions that benefit everyone. In addition, the exchange of ideas is a central component of learning and education.

In a community, people with different backgrounds and experiences come together to share their knowledge and perspectives. This diversity of thought and experience allows for a more holistic and well-rounded learning experience. As individuals learn from each other, they also contribute to the collective wisdom of the community, creating a powerful network of knowledge and understanding.

The sharing of cultural narratives and social ideas has a significant impact on the development and growth of communities. It promotes civic engagement and guides learning and education, ultimately leading to positive change and progress. As individuals continue to exchange ideas and engage with each other, we can create stronger and more resilient communities that work towards a better future for all.

Social cohesion is vital to the development of communities as it strengthens bonds and encourages collaboration. By sharing ideas, individuals build connections and empathy, leading to better relationships and a more harmonious society. As a result, communities become more resilient and united, working towards common goals and creating a better future for all. In addition, civic engagement can promote learning and education through the exchange of ideas. When individuals actively participate in their communities, they gain valuable knowledge and skills while contributing to the growth of others. This continuous learning and sharing can lead to positive change and progress within the community, creating a ripple effect that impacts society as a whole.

As well as promoting social cohesion, the exchange of ideas within a community also promotes adaptability and resilience. When individuals come together to share their thoughts and experiences, they create a collective imagination that can serve as a source of strength in times of uncertainty and adversity. By being open to new ideas and perspectives, communities can more effectively address challenges, embrace change and envision a more resilient future. Furthermore, the act of sharing ideas can stimulate learning and education within a community. Through open and constructive discussion, individuals can learn from each other and increase their knowledge and understanding. This benefits not only the individual, but also the community as a whole, as it becomes more informed and knowledgeable.

By continually sharing and exchanging ideas, communities can foster a culture of learning and education that leads to positive change and progress. In essence, the growth and development of communities is greatly influenced by the act of sharing ideas. Not only does it promote social cohesion, but it also promotes learning and education, which ultimately leads to positive change and progress. By sharing and exchanging ideas, individuals can create stronger and more resilient communities that work towards a better future for all.

The act of exchanging ideas is a fundamental aspect of human interaction and communication. It is a powerful tool for fostering social cohesion and unity within a community. By sharing and exchanging ideas, individuals can better understand each other and develop a sense of empathy and mutual respect. This in turn strengthens the bond between community members and fosters a sense of belonging and togetherness.

Through open dialogue and the sharing of different perspectives and knowledge, individuals can broaden their understanding of the world and expand their horizons. This constant flow of ideas and information leads to personal growth and development, as well as the development of the community as a whole. By continually learning from each other, individuals can work towards creating a more knowledgeable and informed society.

Ultimately, the act of sharing ideas leads to positive change and progress within a community. When individuals come together and share their thoughts and ideas, they can collectively identify and address issues and challenges facing the community. This collaboration and collective problem-solving can lead to innovative solutions and initiatives that benefit everyone.

Individual Emotions

Individual Emotions Individual emotions, often referred to as basic or primary emotions, are rooted in personal experience and physiological responses. They are relatively immediate and direct responses to events that affect the individual. Examples include happiness, sadness, anger, fear, surprise and disgust. While these emotions are influenced by context, they are primarily driven by an individual's evaluation of a situation that affects his or her own well-being or immediate environment. Social emotions, on the other hand, are inextricably linked to our social world. They are shaped by social interactions, cultural norms and our understanding of ourselves in a social context. These emotions often involve a sense of self-evaluation in relation to others or broader social standards. Guilt, shame, pride, envy, empathy and embarrassment all fall under this umbrella. The experience of shame, for example, arises not only from a negative event but also from the perceived judgement or disapproval of others.

Unlike the immediate pain of a physical injury, shame involves a complex process of self-reflection in which one considers one's actions in the light of social expectations and the potential impact on one's social standing. A person may feel ashamed for making a public mistake, not only for the mistake itself, but also for the perceived judgement of others who witnessed it. The emotion is fundamentally intertwined with the social context and the individual's position within it.

The distinction between individual and social emotions isn't always clear-cut; in fact, they often intertwine and influence each other. Consider the experience of fear. While the immediate physiological response to a perceived threat (a speeding car, a sudden loud noise) is an individual emotion, the fear associated with public speaking, for example, is rooted in social anxiety, a concern about the evaluation of others and the potential for social repercussions. In this case, the individual emotion of fear is significantly amplified and shaped by social anxieties and learned associations. The line between individual and social emotions becomes blurred, reflecting the inseparable nature of human experience.

The appraisal process plays a crucial role in distinguishing these emotional states. Cognitive appraisal theory suggests that emotions are not direct responses to stimuli, but rather evaluations of the meaning of events. This appraisal is influenced by personal beliefs, values, past experiences and the social context. An individual's interpretation of an event significantly shapes his or her emotional response.

Cultural contexts have a significant impact on the expression and interpretation of both individual and social emotions. While basic emotions such as happiness and sadness are experienced universally, their expression varies significantly across cultures. Some cultures express emotions openly, while others prefer emotional restraint. These differences are shaped by cultural display rules, learned norms that dictate which emotions are considered appropriate to express in different situations. For example, a public display of grief may be socially acceptable in some cultures but considered inappropriate in others. These display rules have a significant impact on how individuals regulate and express their emotions, particularly social emotions where the social context is paramount.

Even the physiological responses associated with individual emotions can be subtly modulated by culture. Consider the expression of anger. While the physiological arousal associated with anger - increased heart rate, muscle tension - may be similar across cultures, the behavioural manifestations can differ dramatically. In some cultures, open displays of anger are acceptable, even encouraged, while in others anger is suppressed and expressed through indirect means. This cultural shaping highlights the dynamic interplay between biology and culture in shaping emotional experience.

The concept of individual and social emotions goes beyond simple categorisation. It's important to appreciate the intricate interplay between these emotions and their combined impact on human behaviour and social interaction.

The emotional landscape of human experience is far more nuanced than a simple dichotomy would suggest. The universality of basic emotions is a major theme in cross-cultural psychology. While the expression of emotions may vary widely, research suggests that the basic experiences of happiness, sadness, anger, fear and surprise are universally recognised and understood. This universality suggests a biological basis for these emotions, reflecting their fundamental role in human survival and adaptation. However, the experience and interpretation of these emotions is strongly influenced by social and cultural factors.

Furthermore, the development of social emotions is closely linked to cognitive development and socialisation. Self-conscious emotions such as guilt, shame and pride emerge later in childhood and develop alongside an increasing awareness of the self and the social world. These emotions are fundamentally shaped by the feedback and social interactions an individual receives from his or her environment. Understanding these social dynamics plays a crucial role in emotional development.

Understanding the differences and interactions between individual and social emotions provides valuable insights into human behaviour. It allows us to analyse the complexity of social interactions, to understand the motivations behind human actions, and to appreciate the significant role that emotions play in shaping individual lives and societies as a whole. This understanding is not merely an academic pursuit, but an essential tool for navigating the complexities of human relationships and fostering positive social interactions. It underpins the development of emotional intelligence, a key skill for building strong relationships and thriving in social contexts.

Understanding the biological underpinnings of emotions is critical to providing a complete picture, bridging the gap between subjective experience and objective physiological processes.

The amygdala, a small almond-shaped structure deep in the temporal lobe of the brain, plays a central role in processing fear and other related emotions. Numerous studies using functional magnetic resonance imaging (fMRI) and other neuroimaging techniques have consistently shown increased amygdala activity in response to threatening stimuli.

However, the role of the amygdala is not simply a hard-wired response to stimuli. Its activity is modulated by higher order cognitive processes, primarily involving the prefrontal cortex (PFC). Located at the front of the brain, the PFC is responsible for executive functions such as planning, decision-making and working memory. It also plays a crucial role in regulating emotions, acting as a sort of 'brake' on the more impulsive responses of the amygdala. The PFC helps us to contextualise our emotional experiences, taking into account the specific situation and the social implications of our emotional responses.

The interaction between the amygdala and the PFC is not a simple on/off switch, but a dynamic interplay that constantly adjusts the emotional response based on cognitive evaluation. This intricate relationship highlights the complexity of emotion regulation and the brain's remarkable ability to adapt to different situations. Neuroimaging studies have shown that individuals with effective emotion regulation skills have stronger PFC activity and a better balance between amygdala and PFC activation than those with less effective regulation strategies. Looking beyond individual emotions, the neurobiology of social emotions presents a more complex picture. While the amygdala still plays an important role, particularly in processing emotional cues from others, the neural networks involved are much broader and include regions associated with social cognition, empathy and self-awareness. For example, studies of empathy have implicated the insula, a brain region involved in processing bodily sensations and subjective feelings, and the medial prefrontal cortex (mPFC), an area involved in mentalising - the ability to understand the mental states of others.

Neuroimaging studies of social emotions such as guilt and shame show the involvement of additional brain regions, including the anterior cingulate cortex (ACC), which is involved in error detection and conflict monitoring. Experiencing guilt or shame often involves a perceived violation of social norms or expectations, leading to an internal conflict between one's actions and one's desired self-image. The role of the ACC in detecting such conflicts is crucial to the subjective experience of these social emotions.

The intricate interplay of brain regions involved in social emotions underscores their complex nature. Unlike the relatively straightforward neural pathways associated with basic individual emotions such as fear, social emotions involve more extensive and interconnected neural networks, reflecting their intricate relationship to social context, self-awareness and social cognition.

Neurotransmitter systems also play a crucial role in shaping emotional experience. Serotonin, for example, is often associated with mood regulation and is implicated in depression, a state characterised by persistent negative emotions. Dopamine, on the other hand, is involved in reward and pleasure, contributing to positive emotions and motivation. Other neurotransmitters, such as norepinephrine and cortisol, are involved in stress responses and the experience of fear and anxiety. These neurotransmitter systems are not isolated entities; they interact dynamically, influencing each other and shaping the overall emotional experience.

In addition, the effects of hormones cannot be ignored. Oxytocin, often referred to as the 'love hormone', plays a crucial role in social bonding and attachment, and significantly influences social emotions such as empathy and trust. Cortisol, a stress hormone, influences our perception of threat and can affect both individual and social emotional experiences by exacerbating negative emotional responses and impairing social cognitive functions.

Variations in the structure and function of these brain regions and neurotransmitter systems may contribute to individual differences in emotional experience and regulation. Genetic predisposition, environmental factors and life experiences all influence the development and functioning of these neural systems, contributing to a wide range of individual differences in emotional expression and regulation. These differences are crucial considerations when assessing emotional well-being and developing targeted interventions for emotional difficulties.

In addition, research into the neurobiological basis of emotions is ongoing. Advances in neuroimaging techniques and our understanding of neural networks continue to refine our understanding of the complex interplay between the brain and emotions. The integration of different research methodologies, including genetics, neurochemistry and behavioural studies, is providing a more comprehensive and nuanced perspective on the intricate relationship between biology and emotional experience. The future holds the promise of even more precise insights into the neural mechanisms underlying both individual and social emotions, paving the way for more effective interventions and treatments for emotional disorders.

The cognitive appraisal of emotions can profoundly shape our emotional experience. While brain structure and neurochemistry provide the physiological basis for emotion, it is our interpretation of events, rather than the events themselves, that ultimately determines our emotional response. This process of interpretation is known as cognitive appraisal. Cognitive appraisal theory suggests that emotions are not direct responses to stimuli, but rather the result of a cognitive evaluation of those stimuli. This means that the same event can evoke very different emotional responses depending on how it is perceived and interpreted by the individual.

The process of cognitive appraisal is not a passive reception of information; it is an active and constructive process that involves several steps. First, we encounter a stimulus, an event or a situation that demands our attention. This can be anything from a minor inconvenience to a major life event. Next, we evaluate the importance of this stimulus. This assessment involves several dimensions, including the relevance of the event to our personal goals and values, the potential consequences of the event, and the coping resources available to us.

In addition, the assessment process is highly personal and subjective. Our past experiences, beliefs, values and cultural background all influence how we interpret events. An individual with a history of negative experiences may be more likely to interpret ambiguous events negatively, leading to increased anxiety and negativity bias. Conversely, individuals with a more optimistic outlook and strong coping mechanisms may be more likely to reinterpret negative events in a more positive light.

The role of cognitive appraisal is not limited to individual emotions; it also has a significant impact on our experience of social emotions. Consider the emotion of guilt. Guilt arises from the perceived violation of a moral standard, an internalised social norm or an expectation. However, the intensity of the guilt experienced depends on the individual's cognitive appraisal of the situation.

Shame has a similar dynamic. Unlike guilt, shame involves a negative self-evaluation. An individual experiencing shame may perceive him or herself as fundamentally flawed or defective as a result of a perceived failure to meet social expectations. The severity of the shame experienced depends on the individual's sense of self-worth and whether the perceived transgression was irreparable.

Several cognitive biases have a significant impact on our emotional appraisal process. Confirmation bias, for example, refers to our tendency to favour information that supports our pre-existing beliefs, while ignoring contradictory evidence. This can lead to the reinforcement of negative emotional states by focusing on evidence that confirms our fears or anxieties, even if other evidence contradicts this interpretation.

Negativity bias, another common cognitive bias, highlights our tendency to weigh negative information more heavily than positive information. This predisposition often leads to increased threat sensitivity and heightened emotional responses to negative experiences, even when the positive aspects of a situation greatly outweigh the negative. This bias can contribute to a more negative and emotionally intense experience of everyday life, especially when combined with other biases. The impact of these biases on emotional appraisal underscores the subjective nature of emotional experience and the importance of cultivating self-awareness and developing critical thinking skills.

The impact of cognitive appraisal extends beyond the realm of individual experience and has significant implications for social interactions. Our interpretations of the behaviour and intentions of others strongly influence our reactions to them. Misinterpretations, often fuelled by cognitive biases, can lead to misunderstandings and conflict. For example, a perceived slight may trigger feelings of anger and resentment if interpreted as a deliberate act of hostility, whereas the same act may be easily forgiven if interpreted as unintentional.

Effective management of our emotions requires the development of cognitive reappraisal skills. Cognitive reappraisal involves consciously reinterpreting a situation to reduce its emotional impact. It involves challenging negative or distorted thoughts and replacing them with more balanced and realistic perspectives. For example, a person feeling anxious about a public speaking engagement might use cognitive reappraisal to reframe their anxiety as excitement and enthusiasm, focusing on the positive aspects of the event. This conscious shift in perspective can significantly reduce the negative emotional experience.

A full understanding of emotions requires recognition of the profound influence of culture. Culture, which encompasses shared beliefs, values, practices and norms within a group, profoundly shapes how we experience, express and interpret both individual and social emotions. This influence is not merely superficial; it penetrates deep into the fabric of our emotional lives, shaping our emotional repertoires and influencing how we regulate our emotions in social contexts.

An important aspect of cultural influence is the variation in the rules of emotional display. These unwritten social norms dictate which emotions are considered appropriate to express in certain situations and the intensity with which these emotions should be displayed. Cultures differ dramatically in their rules of display. This difference isn't just about outward behaviour; it reflects a fundamental divergence in cultural values and beliefs about emotional expression. Western cultures, often characterised by individualism, emphasise personal autonomy and freedom of self-expression. Accordingly, emotional authenticity is often prioritised, even if it means risking social discomfort. In contrast, many Eastern cultures, characterised by collectivism, emphasise social harmony and group cohesion. Therefore, maintaining social order and avoiding behaviours that might disrupt group harmony take precedence over individual emotional expression. The emphasis here is on emotional regulation, avoiding emotional displays that might be seen as disruptive or disrespectful.

This difference in display rules translates into differences in emotional socialisation. From an early age, individuals are taught specific emotional norms, learning which emotions to suppress, amplify or modify depending on the social context. Children in collectivist cultures are often taught to put the needs and feelings of the group before their own, emphasising emotional self-control and empathy for others. They learn to mask negative emotions such as anger or frustration and focus on maintaining harmony within the family and community. In contrast, children in individualistic cultures are often encouraged to express a wider range of emotions more openly, potentially fostering greater awareness of personal feelings and a stronger sense of self.

The impact of cultural display rules goes beyond simply managing the outward expression of emotions. It significantly shapes the experience of emotion itself. If a culture discourages the outward display of sadness, individuals growing up in that culture may learn to suppress their sadness, potentially leading to internalised emotional distress. They may become less adept at identifying and labelling their sad feelings, which may hinder their ability to effectively manage their emotional state. Conversely, cultures that encourage the open expression of sadness may foster greater emotional awareness and a more robust capacity for emotional processing.

These cultural differences are not confined to sadness. Consider anger. In some cultures, expressing anger openly is seen as assertive and even necessary for conflict resolution. In other cultures, however, open expression of anger is seen as taboo, which can lead to conflict avoidance or recourse to indirect, passive-aggressive forms of anger expression. The consequences of these cultural approaches to anger vary; one may encourage direct conflict resolution, while the other may lead to simmering resentment and strained relationships.

Cultural differences also extend to the interpretation of emotional expressions. A facial expression that might be interpreted as sadness in one culture might be interpreted as something else entirely in another. What is seen as a respectful show of deference in one culture may be interpreted as a sign of weakness or submissiveness in another. These differences in interpretation can lead to cross-cultural misunderstandings and miscommunication, highlighting the crucial role of cultural context in understanding emotional behaviour.

In addition, cultural contexts significantly influence the way we understand and experience social emotions. Social emotions such as guilt, shame, pride and embarrassment are deeply intertwined with social norms and expectations and are therefore subject to considerable cultural variation. For example, the experience of shame varies considerably across cultures. In some cultures, shame is closely associated with maintaining family honour and reputation, whereas in others it may be more closely associated with individual morality. The social consequences of experiencing shame - such as social isolation or ostracism - are also culturally influenced.

Consider the concept of pride. While often associated with positive self-evaluation, the expression and interpretation of pride varies dramatically across cultures. In individualistic cultures, pride is often seen as a positive emotion, reflecting achievement and self-reliance. Open displays of pride may be seen as acceptable, even encouraged. In collectivist cultures, however, overt displays of pride may be seen as arrogant or boastful, potentially undermining social harmony. The emphasis in such cultures may be on humility and modesty, with collective achievements valued over individual recognition.

The role of religion and spirituality is also important in shaping emotional experiences. Religious beliefs and practices often provide a framework for understanding and managing emotions. For example, some religious traditions emphasise stoicism and emotional control, teaching individuals to accept suffering and find meaning in adversity. Other religious traditions encourage the open expression of emotions in prayer or worship, seeing emotional expression as a way of connecting with the divine. Such religious and spiritual influences profoundly shape the emotional landscape of individuals and communities.

This exploration of cultural influences highlights the complex interplay between individual experience and societal norms. While our innate biology and cognitive processes undoubtedly play a significant role in shaping our emotions, cultural factors fundamentally shape the way we express, interpret and even experience these emotions. Understanding these cultural nuances is crucial for effective intercultural communication, empathy and navigating the world's diverse emotional landscapes. Intercultural awareness is becoming not just an interesting academic pursuit, but a critical skill for successful interaction in an increasingly globalised world. Ignoring this critical element leads to misunderstandings, misinterpretations and potentially strained relationships, underscoring the vital need to appreciate the profound influence of culture on the human experience of emotions.

The importance of recognising and respecting these different cultural emotional norms and expressions cannot be overstated, as it has a direct impact on the quality of interpersonal relationships, the effectiveness of communication and ultimately the well-being of individuals and communities.

Emotional development in infancy marks the first emergence of basic individual emotions. Newborns show rudimentary expressions of distress, contentment and interest, reflecting innate biological mechanisms. These early emotional expressions serve important adaptive functions, signalling needs and preferences to caregivers. As babies grow, their emotional repertoire expands. By six months, they typically show clearer expressions of happiness, sadness, anger and fear, reflecting increasingly sophisticated cognitive abilities to discriminate stimuli and associate them with specific emotional responses. These emotions are primarily focused on immediate experience, reflecting a self-centred perspective in which the infant is primarily concerned with his or her own internal state and its interaction with the external environment.

The infant's primary attachment figure plays a crucial role in the development of emotional regulation, providing a secure base from which the infant can explore the world and develop coping mechanisms for distress. The quality of this early attachment has a significant impact on the development of the child's capacity for emotional regulation and social interaction in later life. Secure attachment, characterised by a responsive and consistently available caregiver, promotes emotional security and the development of adaptive coping strategies. Insecure attachment patterns, resulting from inconsistent or unresponsive caregiving, can lead to difficulties in emotional regulation and social adjustment.

Toddlerhood (1-3 years) marks the emergence of more complex emotions, particularly those related to self-awareness. Self-conscious emotions such as shame, pride, guilt and embarrassment begin to develop around the age of two and coincide with the child's increasing understanding of themselves as separate individuals with unique characteristics and identities. These emotions are inextricably linked to social interactions and the child's understanding of societal norms and expectations. Emotional regulation strategies become increasingly sophisticated, moving beyond simple distraction techniques to more advanced strategies such as self-soothing and cognitive reappraisal. Language development plays a crucial role in this emotional growth, providing children with the means to label and express their feelings more effectively.

Middle childhood (6-12 years) is characterised by increasing emotional self-awareness and the ability to manage complex emotions. Children learn to inhibit or modify their emotional responses according to social expectations, demonstrating greater emotional regulation skills. Their understanding of emotional complexity grows as they begin to recognise that people can experience multiple emotions simultaneously or have conflicting emotions. They also begin to understand that emotions can be caused by internal factors (thoughts, beliefs) and external factors (environmental events). Peer relationships become increasingly important and influence the child's emotional development and social-emotional learning. The quality of these relationships has a significant impact on their emotional wellbeing and social competence. Children with strong peer support networks tend to have better emotional regulation and social skills than those who experience peer rejection or isolation.

Adolescence (12-18 years) is a time of significant emotional upheaval and change. Hormonal changes, along with social and cognitive development, contribute to increased emotional reactivity and intensity. Adolescents often struggle to cope with the rapid changes in their emotional landscape, leading to periods of emotional volatility. Identity formation plays a crucial role at this stage of development. Adolescents grapple with questions about their identity, values and place in society, which affect their emotional experiences and relationships. The development of emotional intelligence, the ability to understand and manage one's own emotions and those of others, becomes increasingly important at this stage. Peer relationships remain highly influential, with adolescents often relying on peers for social and emotional support.

Young adulthood (18-40 years) is typically characterised by increased emotional stability and greater self-awareness. Individuals develop more effective strategies for coping with stress and resolving emotional conflicts. Intimate relationships and family formation contribute to a deepening of emotional understanding and empathy. Work and career development also have a significant impact on emotional experience. While the majority of people develop healthy emotional regulation strategies, some may continue to struggle with emotional challenges such as anxiety, depression or relationship difficulties. Seeking professional help at this stage can be crucial in addressing these challenges and promoting emotional wellbeing.

Middle adulthood (40-65 years) often involves a reassessment of life goals and priorities, which affects emotional experiences. Individuals may be grappling with issues related to ageing, mortality and the changing dynamics of family relationships. Increased responsibility and concern for family members, especially children and ageing parents, can create unique emotional stressors. Emotional wisdom and emotional regulation skills acquired throughout life are critical to successfully navigating this challenging life stage. For some, the 'midlife crisis' may be a manifestation of the emotional readjustment required to cope with these significant life changes.

Late adulthood (age 65 and beyond) brings new emotional challenges, including adjustment to retirement, loss of loved ones, and declining physical health. While people may experience increased emotional vulnerability, they often develop greater capacity for emotional acceptance and resilience.

Social connections and supportive relationships remain crucial for maintaining emotional wellbeing in late adulthood. Positive emotional regulation skills developed throughout life become particularly important in coping with the challenges of ageing and potential loss. Maintaining active engagement in meaningful activities, strong social relationships, and possibly emotional support from professional counsellors or therapists all contribute to improved emotional well-being in later life.

Erikson's stages of psychosocial development provide a valuable framework for understanding emotional development across the lifespan. Each stage presents unique challenges that influence emotional growth and development. For example, successfully navigating the identity and role confusion of adolescence contributes to a stronger sense of self and improved emotional stability. Similarly, successfully navigating intimacy versus isolation in young adulthood fosters the capacity for deep and meaningful relationships, contributing to overall emotional well-being.

Attachment theory further illuminates the impact of early relationships on emotional development. Secure attachment, established in infancy, lays the foundation for secure emotional regulation, influencing an individual's ability to cope with stress, form healthy relationships and manage emotional challenges throughout life.

Social influence Emotional contagion, the process by which emotions are passed from one individual to another, has a significant impact on group dynamics, shaping collective behaviour and attitudes. It's not just a matter of imitating facial expressions; it's a deeper process involving the unconscious mirroring of physiological and emotional states, influenced by both conscious and unconscious mechanisms. This emotional resonance within a group can have profound consequences, creating a positive and productive atmosphere or fostering negativity and conflict, depending on the initial emotional trigger and the dynamics of the group itself.

One of the key mechanisms underlying emotional contagion is the process of mimicry. People instinctively, and often unconsciously, tend to mirror the postures, expressions and even physiological states of those around them. This mirroring isn't a conscious choice; it's a largely automatic process rooted in our neural circuitry that facilitates empathy and social bonding. Seeing someone smile, for example, can trigger the same neural pathways in the observer, leading to a feeling of happiness or satisfaction, even in the absence of any direct personal experience causing the emotion. This mimicry goes beyond simple facial expressions to include subtle cues such as tone of voice, body language and even breathing patterns. For example, a tense and anxious individual in a meeting may inadvertently communicate their anxiety to others through their body language, leading to a ripple effect of anxiety throughout the group.

The power of emotional contagion is particularly evident in high-pressure or emotionally charged situations. Consider a courtroom. The tension in the air, heightened by the solemnity of the occasion and the emotions of those involved, can quickly spread to bystanders, causing them to feel anxious or even fearful. Similarly, in a sports arena, the collective excitement and jubilation of a cheering crowd can be highly contagious, creating an atmosphere of shared enthusiasm and energy. These situations demonstrate how emotions can transcend individual boundaries and affect a large group of people. The initial emotional state of a few individuals acts as a catalyst, spreading rapidly through the crowd and intensifying the shared emotional experience.

The impact of emotional contagion is not limited to large, impersonal groups; it plays a significant role in smaller, more intimate social settings such as families and workplaces. In a family, a parent's anxiety about a child's health can easily spread to other family members, creating a climate of shared worry and stress.

The effects of emotional contagion go beyond the immediate group. In the context of social movements, emotional contagion plays a crucial role in mobilising support and collective action. The collective anger and outrage experienced by participants in a protest, for example, can spread rapidly through the crowd, increasing determination and the likelihood of collective action. This is evident in historical examples of social uprisings, where the contagious spread of outrage and shared purpose acted as a powerful catalyst for significant societal change. The visual display of anger, the shared chants and slogans, and the collective experience of oppression all contribute to this intensification of emotions, transforming individual acts of dissent into a cohesive movement.

But the emotional contagion is not always positive. Negative emotions such as fear, anger or anxiety can also spread rapidly through groups, potentially leading to undesirable outcomes. In the context of social media, the spread of fear and misinformation can lead to widespread panic or even violence. The rapid spread of emotionally charged content can amplify negative emotions, creating a self-reinforcing cycle of negativity that can overwhelm individual reasoning and critical thinking. This underlines the importance of critical media literacy and responsible content creation.

Susceptibility to emotional contagion varies between individuals and groups. Factors such as personality traits, past experiences and group norms can influence the extent to which an individual is susceptible to emotional influence. For example, individuals with higher levels of empathy or social sensitivity may be more susceptible to emotional contagion because they are more emotionally responsive to the emotional states of others. Similarly, groups characterised by strong social cohesion and shared values are likely to show a higher degree of emotional convergence, i.e. greater similarity in emotional responses. Conversely, groups with weak social ties may show less emotional synchronisation.

Understanding these factors is central to managing the complex dynamics of emotional contagion within groups.

In addition, the context in which interactions occur significantly modulates the effects of emotional contagion. For example, a crowded, noisy environment can amplify the impact of emotions and make individuals more susceptible to emotional influence.

Emotions are not simply internal experiences; they are powerful forces that profoundly shape our social interactions and significantly influence our decision-making processes. Although often viewed as irrational barriers to sound judgement, emotions are an integral part of social cognition, providing valuable information and shaping our preferences and choices. This section explores the complex relationship between emotions and social decision making, illustrating both the beneficial and detrimental effects of emotional involvement.

The influence of emotions on social decisions is multifaceted. Consider the role of fear in risk assessment. While excessive fear can lead to irrational avoidance of even minor risks, a healthy level of fear can prompt us to take precautions and avoid potentially harmful situations. Similarly, anger can be both a constructive and destructive force in social decision-making. Justified anger at an injustice can motivate us to fight for fairness and equality, leading to positive social change. However, uncontrolled anger can cloud judgement and lead to impulsive and aggressive actions that damage relationships and escalate conflict. The ability to regulate anger, to channel it constructively towards effective problem solving, is crucial to achieving positive outcomes in social interactions. This regulation often involves consciously recognising anger, understanding its origins and actively choosing a more measured response.

Positive emotions, such as happiness and hope, can also significantly influence our social decisions. Happy people are more likely to be cooperative, altruistic and willing to compromise. They may be more willing to forgive perceived wrongs and see the best in others. Similarly, hope can provide the motivation to persevere in challenging situations, promoting resilience and persistence in social endeavours. These positive emotions, often associated with increased pro-social behaviour, contribute significantly to positive group dynamics and enhance collective decision-making processes. However, excessive happiness or optimism can lead to unrealistic expectations and a failure to adequately assess risks, potentially resulting in poor decisions. The balance between positive emotional engagement and rational assessment is crucial for effective social decision-making.

Emotional intelligence, defined as the ability to understand and manage one's own emotions and those of others, plays a critical role in navigating the complexities of social decision-making. People with high emotional intelligence are better able to identify and regulate their own emotional responses, preventing impulsive decisions driven by strong emotions. They also have the ability to empathise with others, understanding their perspectives and emotional states, which enhances their ability to build strong relationships and negotiate conflict effectively. This ability to understand and regulate emotions is crucial for effective leadership, collaborative problem solving and conflict resolution.

The influence of emotions on group decision-making is particularly significant. Groupthink, a phenomenon in which the desire for group harmony overrides critical thinking, illustrates the potential pitfalls of emotional influence. If group members suppress dissenting opinions for fear of conflict or rejection, the resulting decision may be suboptimal or even disastrous. This underscores the importance of fostering an environment that values diverse perspectives and encourages open discussion to mitigate the negative influence of group pressure and promote more rational and effective group decisions. Effective leadership in this context involves actively creating space for dissent, encouraging critical thinking and ensuring that all group members feel comfortable expressing their opinions, even if they differ from the majority view.

Emotional biases can significantly distort our social judgements. Confirmation bias leads us to favour information that supports our pre-existing beliefs and to disregard information that contradicts them, even if it is more accurate. This bias is often exacerbated by emotions; strong feelings about a particular issue can make us more resistant to contradictory information, hardening our position and potentially leading to poor decisions. Recognising the existence of such biases and actively seeking out different perspectives can help mitigate their negative influence on social judgement. This may involve consciously seeking out information that challenges our pre-existing beliefs and actively considering alternative viewpoints before reaching a conclusion.

Interpersonal conflict resolution is also significantly influenced by emotions. Anger, resentment and fear can exacerbate conflict and hinder effective communication and compromise. Conversely, empathy, understanding and willingness to compromise can pave the way for constructive conflict resolution. Effective conflict resolution often involves managing one's own emotions, actively listening to the other person's perspective and seeking mutually acceptable solutions. The ability to separate the emotion from the issue at hand is crucial in this process, focusing on the underlying problem rather than getting caught up in the emotional reactions of those involved. Techniques such as active listening, reframing negative statements and identifying common goals can all contribute to productive conflict resolution.

In addition, cultural differences in emotional expression and norms have a significant impact on social decision-making. Cultures that prioritise emotional restraint may favour more cautious and deliberate decision-making, while cultures that encourage emotional expressiveness may exhibit more impulsive or intuitive decision-making processes. Understanding these cultural differences is critical to effective cross-cultural communication and collaboration, avoiding misunderstandings and promoting more respectful and productive interactions. This awareness can be applied to international business, diplomatic relations and any cross-cultural collaboration where emotional expressions can be misinterpreted if not properly contextualised.

Cognitive reappraisal restructuring can lessen the intensity of the anxiety, leading to a more composed and confident presentation. The efficacy of cognitive reappraisal can vary depending on individual differences in cognitive skills and the specific emotional challenge. In contrast to cognitive reappraisal, which focuses on changing the appraisal of a situation, emotional suppression involves attempting to inhibit the outward expression of an emotion. This strategy is often employed when individuals believe expressing their emotions would be inappropriate or socially unacceptable in a given context. However, accumulating suppressed emotions can be detrimental to long-term well-being, potentially contributing to increased stress, anxiety, and even physical health problems. Studies have linked emotional suppression to a range of negative outcomes, including cardiovascular issues and impaired immune function. The act of suppressing emotions requires significant mental effort, creating a cognitive load that can lead to fatigue and reduced cognitive performance. While seemingly effective in the short-term for maintaining social harmony, the long-term consequences often outweigh the perceived benefits.

Expressive suppression, a closely related concept to emotional suppression, focuses on inhibiting the behavioural expression of emotions, while still allowing for the internal experience of the emotion. For example, someone might experience sadness but consciously choose not to cry or openly display their grief in a public setting. This strategy attempts to maintain a level of emotional control without entirely denying the emotional experience. While potentially less harmful than complete suppression of both the feeling and expression, expressive suppression can still lead to difficulties. The internal experience of the emotion remains, and the inability to fully express it can lead to a sense of frustration and emotional distress, impacting overall well-being. Research suggests that expressive suppression, while less detrimental than complete suppression, can also have negative repercussions on psychological and physical health, albeit potentially less severe.

Beyond cognitive reappraisal and suppression, individuals employ various coping mechanisms to manage their emotions in social situations. Some individuals might engage in problem-focused coping, actively addressing the source of the emotional distress. Other individuals might resort to emotion-focused coping strategies, targeting the emotional response rather than the problem itself. This might include engaging in relaxation techniques, seeking social support, or distracting themselves from the upsetting situation. The effectiveness of each coping mechanism depends on factors such as the nature of the emotional challenge, the individual's personality, and the available social support. Individuals with strong social networks tend to utilize social support more effectively, benefiting from the emotional regulation provided by empathetic interactions with others.

The choice of coping mechanism can significantly impact an individual's social adjustment. Problem-focused coping, when feasible, tends to lead to more effective problem resolution and improved well-being. However, when the problem is intractable, emotion-focused coping strategies can help manage the emotional burden and prevent emotional overwhelm. The inability to effectively regulate emotions can negatively impact social relationships, leading to conflict, isolation, and difficulties in maintaining close bonds. Individuals struggling with emotional deregulation may display erratic behaviour, unpredictable emotional responses, and difficulty empathizing with others, all of which can severely strain their social connections.

Emotional regulation plays an essential role in navigating interpersonal conflicts. Individuals skilled in emotional regulation are better equipped to manage their own emotional responses during conflict, preventing escalation and promoting constructive communication. They can more readily understand and empathize with the other person's perspective, which is essential for achieving mutually acceptable resolutions. Conversely, individuals with poor emotional regulation skills might overreact, become defensive, or engage in aggressive behaviours, hindering productive conflict resolution.

Training in emotional regulation can be highly beneficial in improving interpersonal relationships and conflict management skills. This training often focuses on enhancing self-awareness, improving communication skills, and learning effective coping strategies for dealing with conflict-inducing situations.

Furthermore, emotional regulation is closely linked to self-esteem and self-efficacy. Individuals who feel confident in their ability to manage their emotions are more likely to have higher self-esteem and a stronger sense of self-efficacy. This positive self-perception contributes to greater resilience in the face of social challenges and enhanced social interactions. Conversely, individuals struggling with emotional regulation may experience low self-esteem and reduced self-efficacy, impacting their confidence in social situations and potentially leading to social anxiety or withdrawal.

Cultural factors also significantly influence the expression and regulation of emotions. Some cultures emphasize emotional expressiveness, while others prioritize emotional restraint.

Understanding these cultural differences is essential for effective cross-cultural communication and interaction. Misinterpretations of emotional expressions can lead to misunderstandings and conflict if cultural norms are not taken into account. Sensitivity to cultural differences in emotional expression is therefore essential for fostering positive social interactions across diverse populations.

Building on the previous discussion of emotional regulation, we now turn to the crucial social-emotional intelligences of empathy and compassion. These skills are not merely pleasant personality traits; they are fundamental building blocks of positive social interaction, contributing significantly to the well-being of individuals and the strength of communities. Empathy, the ability to understand and share the feelings of another, and compassion, the feeling that arises in response to the suffering of others, are deeply intertwined yet distinct constructs.

Empathy is a multifaceted process involving both cognitive and emotional components. The cognitive aspect, often referred to as 'cognitive empathy' or 'perspective taking', involves the ability to understand another person's mental state - their thoughts, feelings and intentions. It involves stepping out of one's own perspective and trying to see the world from another's point of view. This cognitive understanding is crucial for predicting behaviour and responding appropriately in social interactions.

The emotional component of empathy, known as 'emotional empathy' or 'affective empathy', involves sharing the emotions of the other person. This is not just understanding their feelings, but actually feeling them. If we see someone crying, we might feel a twinge of sadness ourselves. This shared emotional experience fosters connection and motivates pro-social behaviour - actions intended to benefit others. The intensity of emotional empathy can vary greatly between individuals, influenced by factors such as temperament, personal experience and neurological factors. People who are high in emotional empathy are often more sensitive to the needs of others and more likely to engage in helping behaviours.

The interplay between cognitive and emotional empathy is complex and not always straightforward. Individuals may have strong cognitive empathy, understanding another's perspective intellectually, without necessarily experiencing a shared emotional response. Conversely, someone may experience a strong emotional response to another's distress without fully understanding the reasons for their feelings. Ideally, both cognitive and emotional empathy work together to create a rich and nuanced understanding of another's experience, fostering deeper connections and more effective social interactions.

The development of empathy is a complex process influenced by both biological and environmental factors. Genetic predispositions play a role, but early childhood experiences are particularly critical. Secure attachment relationships with caregivers, characterised by consistent responsiveness and emotional support, are associated with the development of higher levels of empathy. Children who experience consistent care and understanding from their caregivers are more likely to develop the ability to understand and share the emotions of others. Conversely, childhood experiences of neglect, abuse or trauma can have a negative impact on the development of empathy, making it more difficult to relate to and understand the feelings of others. Parenting styles that promote perspective taking, emotional expression and regulation have a significant impact on the development of empathy in children.

The expression and experience of empathy is not static; it is strongly influenced by social and cultural factors. Cultural norms influence how emotions are expressed and interpreted, and affect the way empathy is displayed and experienced in different societies. Some cultures may emphasise emotional restraint, making it less common to overtly express empathy through emotional displays. Others may encourage open displays of emotion, creating a more emotionally expressive environment in which empathy is more readily apparent. Understanding these cultural differences is critical to avoiding misinterpretations and promoting positive interactions across cultures.

Individual differences in personality traits also influence the expression and experience of empathy. Individuals high in agreeableness and openness to experience tend to show higher levels of empathy than those low in these traits. However, it's important to understand that empathy is not simply a matter of personality; it's a skill that can be cultivated and strengthened through conscious effort and practice. Developing skills in perspective taking, active listening and emotional regulation can enhance one's capacity for empathy. Mindfulness practices can also help cultivate empathy by improving self-awareness and the ability to focus on the experiences of others.

Empathy plays a crucial role in various aspects of social life, significantly influencing interpersonal relationships, leadership styles and conflict resolution. In interpersonal relationships, empathy is a cornerstone of intimacy and trust. People who demonstrate empathy are better able to understand their partner's needs and concerns, leading to stronger and more satisfying relationships. They are more attuned to their partner's emotional cues, fostering a sense of emotional safety and connection. Research consistently shows a strong link between empathy and relationship satisfaction across different types of relationships, including romantic partnerships, friendships and family relationships.

In leadership, empathy is increasingly recognised as a critical leadership skill. Empathetic leaders are better able to understand the needs and motivations of their team members, fostering a supportive and collaborative working environment. They can identify and address sources of stress and conflict, leading to increased productivity and job satisfaction among employees. Empathetic leaders demonstrate a genuine concern for the well-being of their team, creating a positive and motivating work culture. This is particularly important in today's diverse and complex workplaces, where effective communication and understanding are essential.

Empathy also plays an important role in conflict resolution. People who can empathise with the other party's perspective are better able to understand their needs and concerns, making it more likely that a mutually acceptable solution can be reached. Empathy enables individuals to see the situation from multiple perspectives, reducing the likelihood of conflict escalation. By demonstrating understanding and acknowledging the other person's feelings, empathetic individuals create a safer space for communication and negotiation. This is true in both personal and professional contexts, where conflict resolution is a critical skill for maintaining positive relationships and achieving collaborative outcomes.

The concept of compassion builds on empathy, adding a sense of concern and motivation to alleviate the suffering of others. Compassion is not simply feeling the pain of others; it involves a desire to act on those feelings by taking steps to provide help and support. Compassionate people are motivated to alleviate the suffering of others, whether through direct action or advocacy. This active element distinguishes compassion from empathy, which goes beyond understanding another's feelings to actively engaging in helping behaviour.

Compassion is also a skill that can be cultivated and improved. Practices such as mindfulness meditation have been shown to increase levels of compassion. These practices cultivate self-awareness and empathy, creating a foundation for compassionate action. In addition, engaging in acts of kindness and service to others can enhance one's capacity for compassion. By experiencing the positive emotions associated with helping others, individuals become more likely to engage in future acts of compassion. This virtuous cycle reinforces the development of compassionate responses.

Cultivating empathy and compassion is not only beneficial for individuals, it is also crucial for fostering strong and supportive communities. Societies characterised by high levels of empathy and compassion tend to be more equitable and just. Empathetic individuals are more likely to support social justice initiatives, recognising the suffering and injustice experienced by marginalised groups. They are also more likely to engage in pro-social behaviours that contribute to the well-being of their communities. Cultivating empathy and compassion within communities is a critical step in building more resilient, inclusive and harmonious societies.

The cultivation of empathy and compassion provides a crucial foundation for the development of robust social support networks. These networks, made up of individuals who provide emotional, informational or practical assistance, act as a critical buffer against the negative effects of stress and adversity. The strength and quality of these networks significantly influence an individual's ability to regulate emotions and cope effectively with life's challenges. Individuals who are embedded in strong social networks show an increased ability to cope with stressful situations and exhibit greater resilience and emotional well-being than those who experience social isolation.

The protective effect of social support manifests itself in a number of ways

People with strong social ties often experience lower levels of the stress hormone cortisol in response to stressful events. This physiological response reflects the calming and supportive nature of their social connections, which reduces the physiological burden of stress. In addition, social support is a source of emotional regulation, providing a safe space for individuals to express their feelings without judgement. The act of sharing emotional burdens with trusted others can significantly reduce the intensity of negative emotions, preventing them from escalating into more serious mental health problems. This emotional release is facilitated by the empathy and compassion of supportive individuals, which validates the individual's experience and reduces feelings of isolation and vulnerability.

The provision of practical assistance further enhances the protective role of social support. In times of need, help from others, ranging from childcare to financial assistance, can significantly reduce the burden of stressors, lowering overall stress levels and improving emotional well-being. This practical support directly reduces the individual's stress load, freeing up mental and emotional resources to focus on coping and recovery. This form of support is particularly important at times of major life transitions, such as job loss, bereavement or serious illness.

Beyond the immediate benefits, strong social networks contribute to a broader sense of belonging and purpose, fostering an enhanced sense of self-worth and identity. This enhanced sense of self acts as a protective factor against the negative effects of stress, enabling individuals to face challenges with greater confidence and resilience. Knowing that they have a network of supportive people who care about their wellbeing increases their self-efficacy and reduces feelings of helplessness and despair. This sense of belonging is particularly important in times of transition or uncertainty, providing a stable anchor in the midst of life's fluctuations.

Conversely, the absence of strong social support, manifested as social isolation or loneliness, poses significant risks to mental and physical health. Chronic loneliness, characterised by persistent feelings of isolation and disconnection, has been consistently linked to an increased risk of depression, anxiety, cardiovascular disease and even premature mortality. The detrimental effects of loneliness underscore the fundamental human need for connection and belonging. The lack of social support leaves individuals vulnerable to the overwhelming effects of stress, without the crucial buffer of supportive relationships. Without a network of trusted individuals to confide in and share burdens with, individuals struggling with stress are more likely to experience heightened negative emotions and impaired coping mechanisms. This isolation amplifies the negative effects of stressors, contributing to a vicious cycle of distress and further withdrawal from social interaction.

The detrimental effects of social isolation are particularly pronounced in vulnerable populations, such as older people, people with disabilities and people with mental health problems. These groups may face greater barriers to accessing social support, further exacerbating the negative consequences of loneliness. The development of effective interventions to reduce social isolation and promote social connectedness in these populations is a critical public health imperative. Such interventions may include community programmes that promote social engagement, technological solutions that facilitate virtual connections, and initiatives that address barriers to social participation, such as transportation or accessibility issues.

The impact of social support extends beyond individual wellbeing, significantly influencing community resilience and societal stability. Strong social networks enhance the ability of communities to respond effectively to collective stressors, such as natural disasters or economic downturns. The mutual support and cooperation fostered by strong social ties enable communities to cope with challenges more effectively, providing essential resources and emotional support during difficult times. This collective resilience is underpinned by the ability of communities to use social capital effectively, translating social connections into tangible benefits for the collective good.

Research on social support and emotional well-being highlights the crucial interplay between individual and social factors in influencing emotional experience and coping. While individual emotional regulation skills are undeniably important, the availability and quality of social support significantly modulate the impact of stressors on emotional well-being. Investing in strengthening social networks and reducing social isolation is crucial to promoting individual and collective resilience and creating healthier and more supportive communities. This requires a multi-faceted approach that includes individual-level interventions that promote social skills and emotional intelligence, as well as community-based programmes that promote social connectedness and reduce social inequalities, ensuring that people from all backgrounds have access to the vital social support they need to thrive.

Further research could explore the effectiveness of different interventions aimed at strengthening social support networks, such as peer support groups, community-based mental health programmes and technological solutions that promote social connectedness. These studies should consider the specific needs and challenges of different populations and tailor interventions to maximise their effectiveness and accessibility. Understanding the mechanisms by which social support exerts its protective effects is also a critical area for future research, potentially revealing new pathways for improving emotional well-being. This research should also examine the role of cultural context in shaping both the provision and receipt of social support, highlighting the need for culturally sensitive interventions that are consistent with community norms and values.

Group dynamics, the patterns of interaction and relationships within a group, are profoundly influenced by the emotional climate. This climate is not simply the aggregate of individual emotions;

it's a shared, emergent property shaped by the interplay of individual feelings, group norms and situational factors. A positive emotional climate, characterised by feelings of trust, optimism and mutual respect, encourages collaboration, innovation and high performance.

Members are more likely to communicate openly, participate actively in decision making and support each other's efforts. This positive feedback loop strengthens group cohesion and facilitates the achievement of collective goals. For example, a research team characterised by a positive emotional climate, where members feel comfortable sharing ideas and giving constructive feedback, is more likely to produce ground-breaking research than one characterised by negativity and mistrust.

Conversely, a negative emotional climate, characterised by feelings of fear, hostility and mistrust, can significantly impair group functioning. In such environments, communication breaks down, conflict escalates and individual performance suffers. Members may become withdrawn, reluctant to share ideas or seek help, leading to reduced creativity and productivity. The constant presence of negative emotions can also lead to burnout, reduced job satisfaction and increased absenteeism. Consider, for example, a workplace characterised by high levels of stress, competition and a lack of supportive leadership. Employees in this environment are more likely to experience emotional exhaustion, cynicism and reduced professional efficacy, leading to reduced productivity and increased turnover.

The concept of emotional contagion plays an important role in shaping the emotional climate of a group. Emotional contagion is the process by which emotions spread from one person to another, affecting the overall emotional state of the group. Positive emotions, such as enthusiasm and joy, are highly contagious, creating a ripple effect that boosts group morale and motivation. Conversely, negative emotions such as anger and fear can spread quickly, creating a climate of tension and anxiety that inhibits effective collaboration. A single negative interaction or outburst can trigger a cascade of negative emotions, disrupting group dynamics and undermining performance. This phenomenon is particularly relevant in organisations where the emotional state of a leader can significantly influence the overall emotional climate of the team. A leader who displays negativity or hostility can unintentionally spread these emotions throughout the organisation, negatively impacting employee morale, productivity and well-being.

Group norms, the unwritten rules that govern behaviour within a group, also have a significant impact on emotional climate. Groups with norms that encourage emotional expression and open communication tend to have more positive emotional climates. Members feel comfortable sharing their feelings, offering support and resolving conflicts constructively. In contrast, groups with norms that suppress emotional expression or discourage open communication tend to have negative emotional climates. Members may bottle up their feelings, leading to resentment, frustration and conflict. These norms may be implicit, learned through observation and social learning, or explicit, defined by formal policies or statements.

The impact of emotional climate goes beyond the immediate functioning of the group, influencing long-term outcomes such as group cohesion, stability and even survival. Groups with a positive emotional climate tend to be more cohesive, with members feeling a strong sense of belonging and shared identity. This cohesion fosters loyalty, commitment and a willingness to work together towards collective goals. In contrast, groups with negative emotional climates are often plagued by conflict, distrust and high turnover, leading to instability and reduced effectiveness. The impact of emotional climate is also visible in the context of organisations.

Organisations with positive emotional climates attract and retain top talent, increasing their overall performance and competitiveness. These organisations also tend to be more innovative and adaptable, better equipped to manage change and overcome challenges.

The influence of emotional climate on group outcomes is supported by extensive research in a variety of fields, including organisational psychology, social psychology and sociology. Studies have shown a strong correlation between a positive emotional climate and improved group performance, increased creativity and better decision-making. These studies have also highlighted the detrimental effects of negative emotional climates, linking them to reduced productivity, increased conflict and increased burnout. Several strategies can be implemented to cultivate a positive emotional climate within groups. These strategies often include encouraging open communication, providing opportunities for emotional expression, and promoting empathy and compassion.

Leaders play a critical role in setting the emotional tone of the group, modelling positive behaviours and creating a supportive environment. Emotional intelligence training programmes can also help members develop skills in emotional regulation, empathy and conflict resolution. The importance of effective leadership cannot be overstated. Leaders who are emotionally intelligent, able to understand and manage their own emotions and empathise with others, are crucial in creating a positive emotional climate. They create a sense of psychological safety where team members feel comfortable speaking up, taking risks and supporting each other. This creates a self-reinforcing cycle of positive emotions that leads to increased collaboration and productivity.

Conflict is an inherent part of group dynamics, arising from differing perspectives, goals, values, or communication styles. While some level of conflict can be constructive, stimulating creativity and problem-solving, unchecked conflict can quickly become detrimental, eroding team cohesion, reducing productivity, and fostering a hostile work environment. The ability to navigate conflict effectively, therefore, is a critical skill not only for leaders but for all members of a social group or organization.

One of the primary challenges in conflict resolution lies in the intense emotions that often accompany disagreements. Anger, frustration, fear, resentment, and defensiveness are common responses to conflict, and these emotions can easily escalate the situation, hindering rational discussion and compromise. Effective conflict resolution necessitates a conscious effort to regulate these emotions, ensuring that they do not derail the process. This involves developing self-awareness—recognizing the emergence of these emotions within oneself—and self-regulation—managing those emotions to prevent impulsive or destructive reactions.

Beyond individual emotional regulation, effective conflict resolution demands empathy and perspective-taking. Understanding the other party's perspective, even if one disagrees with it, is crucial for de-escalating tension and finding common ground. Active listening—paying close attention to what the other person is saying, both verbally and nonverbally—is essential for comprehending their viewpoint. Reflecting back what one has heard, paraphrasing their concerns to ensure understanding, demonstrates empathy and validates their feelings, helping to build trust and foster collaboration.

The ability to separate the person from the problem is another critical aspect of emotional regulation in conflict. Personal attacks and blaming language often escalate conflict, making it difficult to focus on the issue at hand. By focusing on the specific problem or disagreement, rather than making personal judgments about the other individual, participants can maintain a more objective and constructive approach. Using "I" statements, expressing one's own feelings and needs without blaming the other person, is a valuable technique for promoting respectful communication.

Several models for effective conflict resolution provide structured approaches to managing emotions and navigating disagreements. Negotiation, for instance, involves a collaborative process where parties work together to find a mutually acceptable solution. Successful negotiation requires emotional intelligence, the ability to understand and manage one's own emotions, and to empathize with the other party's perspective. Compromise and finding creative solutions often involve managing conflicting emotions and desires.

Mediation, on the other hand, involves a neutral third party who facilitates communication and helps the conflicting parties to reach a resolution. Mediators play a vital role in managing emotions, helping individuals to express their feelings constructively and guiding them towards a mutually agreeable outcome. Their expertise lies in de-escalating tensions, creating a safe space for communication, and guiding parties toward identifying common interests and finding creative solutions. Mediators often employ techniques that emphasize active listening, empathy, and reframing disagreements to focus on shared goals.

The choice between negotiation and mediation depends on the nature of the conflict and the willingness of the parties involved. Simple disagreements might be effectively resolved through negotiation, while more complex or emotionally charged conflicts might benefit from the intervention of a skilled mediator.

In another scenario, a conflict might arise between a manager and an employee regarding performance expectations. A manager who lacks emotional intelligence might respond with anger or frustration, leading to a confrontational and unproductive discussion. However, an emotionally intelligent manager would recognize their own emotions and employ strategies for self-regulation, approaching the conversation with empathy and understanding. Active listening and clear communication could lead to a productive dialogue that helps to clarify expectations and improve the employee's performance. If the conflict persists despite these efforts, a mediation process might be helpful.

Beyond specific techniques, the broader organizational culture significantly influences how conflicts are handled. An organization that values open communication, encourages feedback, and provides training in conflict resolution skills fosters a more positive and constructive approach to disagreements. Creating a psychologically safe environment where individuals feel comfortable expressing their opinions and concerns without fear of retribution is crucial for preventing conflict escalation and promoting effective resolution.

The effectiveness of individual emotional regulation strategies is significantly influenced by the broader organizational context. The impact of organizational culture on emotional expression manifests in several key ways. Some organizations foster cultures that actively encourage emotional expression, viewing open communication and emotional transparency as vital for collaboration and innovation. In these settings, employees feel comfortable sharing their feelings, both positive and negative, with colleagues and supervisors without fear of retribution or judgment.

This open emotional climate can lead to increased trust, stronger team cohesion, and enhanced creativity. Employees feel more psychologically safe, leading to higher levels of engagement and job satisfaction. They are more likely to take risks, voice dissenting opinions, and contribute innovative ideas, knowing their contributions will be received with respect and understanding, even if they challenge the status quo. Feedback mechanisms are often more robust and constructive, facilitating continuous improvement and learning.

In contrast, other organizations cultivate cultures where emotional expression is suppressed or even discouraged. These organizations often prioritize rationality, control, and efficiency above all else. Emotional displays, particularly negative ones like anger or frustration, are viewed as unprofessional, disruptive, or even threatening. In such environments, employees may feel pressured to mask their true feelings, suppressing emotions to conform to organizational norms. This suppression of emotions, however, can have significant negative consequences. Conflict resolution becomes more difficult, as suppressed emotions often manifest in passive-aggressive behaviours or outright explosions, rather than constructive dialogue.

The impact of these differing cultural norms extends beyond individual well-being; it significantly affects organizational performance. Organizations that encourage emotional expression often experience higher levels of innovation, creativity, and problem-solving.

An influential model, Hofstede's cultural dimensions theory, examines cultural variations across national and organizational contexts. One dimension, masculinity versus femininity, highlights the contrast between cultures that prioritize assertiveness, competition, and material success (masculine) versus those that emphasize cooperation, nurturing, and quality of life (feminine). Organizations reflecting a masculine culture might suppress emotional expression, especially emotions associated with vulnerability or weakness, while organizations leaning towards a feminine culture might be more tolerant and even encouraging of emotional displays. Similarly, Hofstede's dimension of power distance – the extent to which less powerful members of organizations accept unequal power distribution – can influence emotional expression.

While economic disparities and power imbalances undeniably shape social inequality, the emotional landscape has a significant impact on how these inequalities manifest and are perpetuated. Negative emotions, particularly fear, anger and disgust, often serve as the emotional underpinnings of prejudice, shaping prejudiced attitudes and discriminatory behaviour towards out-groups. Understanding the emotional mechanisms that underpin prejudice is crucial to dismantling its pervasive influence on social structures.

Fear, a fundamental human emotion, plays an important role in shaping prejudiced attitudes. When individuals perceive an outgroup as a threat, whether real or imagined, fear can trigger a cascade of negative emotions and behaviours. This threat perception, often fuelled by misinformation, stereotypes and societal fears, can lead to the dehumanisation of the outgroup, making it easier to justify discriminatory actions. For example, historical and contemporary examples of xenophobia often illustrate how fear of 'the other' - often rooted in cultural differences, religious beliefs or perceived economic competition - fuels discriminatory policies and practices. Fear of immigrants taking jobs, for example, can lead to restrictive immigration policies and hostile attitudes towards immigrant communities, regardless of the validity of the underlying concerns.

Similarly, anger can be a powerful driver of prejudice. When individuals feel wronged or perceive injustice, anger can be directed at perceived wrongdoers, often leading to the scapegoating of out-groups. This is particularly common in times of economic hardship or social instability. In times of economic recession, for example, anger and frustration may be directed at minority groups, leading to increased incidents of hate crime and discriminatory practices. The scapegoating mechanism allows individuals to displace their anger and frustration onto a convenient target, rather than addressing the underlying systemic issues that contribute to their dissatisfaction. This emotional displacement reinforces existing prejudices and creates a self-fulfilling cycle of anger and discrimination.

Disgust, another emotion with powerful social implications, plays a crucial role in shaping prejudice, particularly towards groups perceived as morally deviant or physically impure. Often associated with feelings of contamination and disgust, disgust can lead to social distancing and discriminatory actions against those perceived as different. This can take many forms, from subtle expressions of avoidance and exclusion to overt acts of aggression and violence. Historical examples of racial discrimination, in which certain groups were seen as inherently 'unclean' or 'inferior', provide a powerful illustration of the destructive potential of disgust-based prejudice. These social constructions of disgust, perpetuated over generations, have created deep-seated prejudices that continue to affect social relations and equality.

However, the emotional landscape is not just made up of negative emotions that fuel prejudice. Positive emotions such as empathy, compassion and love can play a vital role in promoting tolerance, understanding and social cohesion. Empathy, the ability to understand and share the feelings of others, is a powerful antidote to prejudice. When people can put themselves in the shoes of those from different backgrounds, they are more likely to recognise their common humanity and overcome ingrained prejudices. Fostering empathy through education, intercultural exchange and meaningful interaction with people from different backgrounds can significantly reduce the prevalence of prejudice.

Compassion, which is closely related to empathy, involves a sense of concern for others and a desire to alleviate their suffering. When individuals feel compassion for those who are marginalised or disadvantaged, they are more likely to support policies and initiatives that promote social justice and equality.

For example, compassionate leadership can create a workplace environment where diversity is valued and discrimination is actively challenged. Cultivating compassionate leadership within organisations and wider society can contribute to building a more inclusive and just world.

Fostering a sense of collective belonging can challenge existing prejudices and promote tolerance. The interplay between emotions and prejudice is not just a matter of individual attitudes; it is deeply embedded in broader societal structures and cultural norms. Media representations, political rhetoric and social institutions all contribute to shaping emotional responses to outgroups. Negative portrayals of minority groups in the media, for example, can fuel fear and mistrust and exacerbate existing prejudices. Conversely, positive and inclusive media portrayals can promote understanding and empathy, contributing to a more tolerant society. Understanding how these societal mechanisms work is crucial to developing effective strategies to combat prejudice.

Addressing the emotional roots of prejudice requires a multifaceted approach. At the individual level, interventions such as the promotion of emotional intelligence, empathy training and cognitive behavioural therapy can help individuals to recognise and manage their own biases. At the societal level, initiatives to promote intercultural dialogue, media literacy and social justice education can challenge societal norms that perpetuate prejudice. In addition, fostering inclusive environments where people from diverse backgrounds feel safe, respected and valued can create a more tolerant and equitable society. Taken together, these initiatives can facilitate a gradual shift in the emotional landscape, helping to reduce prejudice and promote a more just and harmonious society.

Ultimately, overcoming the emotional barriers to social equality requires a sustained and collaborative effort involving individuals, communities and institutions. By understanding the complex interplay between emotions, prejudice and social inequality, effective strategies can be developed to dismantle the emotional structures that perpetuate injustice and promote a more inclusive and equitable future. This requires a holistic approach that addresses both individual biases and the broader societal forces that reinforce prejudice, focusing on promoting positive emotions and challenging the negative ones that fuel discrimination. Only through such a comprehensive approach can we move towards a world where emotional equality reflects social justice.

Understanding these different emotional responses is essential to understanding the dynamics of social movements, activism and the pursuit of social justice. Anger undoubtedly remains a prominent emotion in the face of injustice. But it is not simply a reactive, individualistic emotion. In the context of social inequality, anger can be a powerful catalyst for collective action. When individuals perceive systemic injustices - whether racial discrimination, economic exploitation, or gender inequality - their individual anger can coalesce into a shared sense of outrage that drives mobilisation and social change. The impassioned speeches of Martin Luther King Jr, reflecting the frustration and anger of a community, served as a rallying cry, inspiring collective action and demanding social justice. This collective anger, channeled through peaceful protests, boycotts and civil disobedience, ultimately played a pivotal role in dismantling segregation and advancing civil rights.

Similarly, the women's suffrage movement, which spanned decades of struggle, was fuelled by the collective anger and frustration of women denied the basic right to vote. The collective emotion of anger, expressed through marches, demonstrations and sustained activism, eventually led to significant legal changes that empowered women and transformed their social standing. These historical examples demonstrate the transformative power of collective anger when harnessed to challenge unjust systems. However, it's important to recognise that anger, if not properly channelled and managed, can also lead to violence and counterproductive outcomes. The critical element is the ability to channel this powerful emotion into constructive social action focused on bringing about meaningful change. This requires careful organisation, leadership and a strategic approach to advocacy and protest.

Sadness, often perceived as a passive emotion, can also play an important role in responding to injustice. The experience of personal or collective loss resulting from discriminatory policies or practices can lead to deep sadness and despair. This sadness can also be a powerful motivator for action. Awareness of widespread suffering and inequality can evoke feelings of empathy and compassion, motivating individuals to work to alleviate the suffering of others. For example, the profound sadness associated with the devastating effects of poverty can fuel social movements to address income inequality and provide social support to vulnerable populations. The emotional resonance of witnessing the pain and hardship of others can be a more compelling call to action than abstract notions of fairness or justice. Moreover, this grief can be a powerful catalyst for personal reflection and engagement in meaningful social work, offering a profound pathway to restorative justice.

Beyond anger and sadness, hope is an important emotional component in the struggle for social justice. Hope, the belief that change is possible and that positive outcomes can be achieved, is essential for maintaining momentum and sustaining collective action over time. Social movements rarely achieve their goals quickly; they often involve long campaigns, setbacks and periods of discouragement. The ability to maintain hope in the face of adversity is crucial to sustaining the energy and commitment necessary to achieve lasting change. The emotional dynamics of social movements aren't determined solely by the experience of injustice; they are also shaped by factors such as leadership styles, organisational structures and the narratives used to frame the movement's goals. Inspirational leaders, who are able to articulate shared grievances and mobilise collective emotions, play a crucial role in shaping the emotional trajectory of a social movement. Their ability to connect with individuals on an emotional level, validating their experiences and feelings, is critical to fostering commitment and sustaining participation. This is particularly important in ensuring that the anger or sadness experienced is effectively channelled into constructive action.

In addition, narrative framing, particularly the stories and symbols used to define the goals and values of the movement, has a significant impact on the emotional responses of participants and bystanders. Framing a movement around values of justice, compassion and human dignity can foster positive emotional responses and broaden the movement's base of support. Conversely, narratives that demonise opponents or use fear-mongering tactics can generate negative emotions and potentially undermine the goals of the movement. Therefore, strategic communication and framing are essential to cultivate the emotional climate necessary for effective mobilisation and long-term success.

The interplay between individual and collective emotions in the context of social injustice is a complex and dynamic phenomenon. While individual emotions can provide the initial impetus for engaging in social action, it's the collective emotional responses that truly drive social movements and shape the course of social change.

The long-term effects of systemic disadvantage require a deeper understanding of emotional resilience and the critical role that social support networks play in fostering it within marginalised communities. Resilience in this context is not simply the absence of negative emotions; it's the ability to cope with adversity, to recover from setbacks, and to maintain a sense of well-being despite ongoing challenges. It's a dynamic process, shaped by both individual characteristics and the supportive environments in which individuals are embedded.

Marginalised communities, facing persistent systemic barriers to economic opportunity, health care, education and justice, often endure a cumulative burden of stress. This chronic stress, resulting from discrimination, prejudice and socio-economic disadvantage, can have a significant impact on mental and physical health. The constant vigilance against micro-aggressions, the pervasive experience of prejudice, and the struggle for basic necessities can lead to an increased risk of anxiety, depression, and post-traumatic stress disorder (PTSD). Yet despite these significant challenges, many people from marginalised communities demonstrate remarkable resilience, the ability to cope effectively with adversity and maintain a positive outlook. This resilience isn't simply an innate trait; it is cultivated and nurtured within the supportive contexts of family, community and cultural institutions.

One of the most important buffers against the damaging effects of marginalisation is the presence of strong social support networks. These networks, formed through family ties, community organisations, religious institutions and peer groups, provide individuals with a sense of belonging, validation and practical assistance. Family structures, often the primary source of social support, can serve as havens of emotional security and practical assistance. Within these family units, individuals can share their experiences, seek emotional comfort and receive practical assistance in overcoming systemic barriers. The strength of family ties can significantly mitigate the impact of stress and adversity and promote a sense of stability and hope. This is particularly important for children growing up in marginalised communities, who may otherwise face significant emotional and psychological challenges.

Beyond the family unit, community organisations play a crucial role in providing social support and fostering resilience. Often rooted in shared ethnicity, religion or socio-economic background, these organisations provide spaces for social interaction, mutual support and collective action. They provide platforms for sharing experiences, building solidarity and developing collective strategies to address systemic challenges. Community centres, religious congregations and mutual aid societies offer not only social connection but also practical support, including access to resources, vocational training and legal aid. This practical support can be particularly important in mitigating the socio-economic inequalities that often exacerbate the negative effects of marginalisation.

Religious institutions serve as vital sources of spiritual solace, emotional support and community cohesion for many individuals within marginalised communities. Religious faith can provide a framework for understanding suffering, finding meaning in adversity, and maintaining hope in the face of systemic injustice. Religious communities provide spaces for social connection, shared worship and mutual support, fostering a sense of belonging and collective identity. Pastors, priests, rabbis and other religious leaders often play a crucial role in providing advice, guidance and practical assistance to members of their congregations. The spiritual and emotional support offered by religious institutions can be particularly important for people who have experienced trauma or are facing significant challenges.

However, it's important to recognise that the effectiveness of social support networks is not consistent across marginalised communities. Factors such as the strength of community ties, the availability of resources and the level of social cohesion can significantly influence the extent to which social support networks can buffer against the negative effects of marginalisation. In communities characterised by high levels of poverty, social fragmentation and limited access to resources, social support networks may be weaker and less effective in providing adequate protection against the cumulative burden of stress. Similarly, in communities experiencing high levels of violence or social unrest, social support systems may be strained and less able to provide the necessary emotional and practical support.

In addition, the nature of social support can vary considerably between communities and cultures. While close-knit family structures may be the primary source of support in some communities, others may rely more heavily on broader community networks or religious institutions. Understanding the specific cultural contexts that shape social support within different marginalised communities is central to developing culturally sensitive interventions to promote resilience and well-being. This includes recognising the different ways in which individuals within these communities cope with adversity and seek support. Cultural factors can shape how emotions are expressed, how social support is sought and provided, and how individuals define resilience in their specific contexts.

Beyond formal support networks, informal social connections also contribute significantly to emotional resilience. Friendships, peer support groups and informal mentoring relationships can provide important sources of emotional validation, practical help and encouragement. These informal connections can be particularly important for people who may not have access to formal support services, or who may prefer to seek support from peers who understand their experiences. Moreover, the sense of belonging and shared identity that comes from informal social connections can be a powerful antidote to the isolation and marginalisation that often accompany systemic disadvantage.

In addition to providing emotional support, social support networks also play a crucial role in facilitating access to resources and opportunities. They can provide assistance with job searches, housing, healthcare and legal aid, thereby reducing the socio-economic inequalities that contribute to stress and adversity. This practical help can be particularly important for people who face barriers to accessing mainstream services or who may be reluctant to seek help from institutions that have historically discriminated against them. The ability of social support networks to provide practical help underlines their importance in reducing stress and promoting resilience within marginalised communities.

However, it's important to recognise the limitations of social support networks. While they can be powerful buffers against the negative effects of marginalisation, they cannot fully compensate for the systemic inequalities that generate stress and adversity.

Addressing the root causes of marginalisation - racism, sexism, economic inequality and other forms of discrimination - is essential to creating truly equitable and just societies. While social support networks are essential for building resilience, they are not a substitute for broader societal changes aimed at addressing systemic inequalities. Social support fosters resilience, but systemic change is needed to address the root causes of the stress these systems create.

Political polarisation Political polarisation, the divergence of political attitudes towards ideological extremes, is not a new phenomenon. However, the current era is characterised by increased intensity and deeper societal divisions, largely fuelled by the amplification of emotional responses through various communication channels, especially social media. This amplification effect transforms simple disagreements into deeply entrenched, emotionally charged conflicts, making compromise and cooperation extremely difficult.

The emotional landscape of political discourse has become increasingly toxic, characterised by hostility, distrust and a pervasive sense of threat. This emotional climate undermines the ability of citizens to engage in rational discourse, leading to political gridlock and a decline in the overall health of democratic institutions.

One of the key drivers of this emotional polarisation is the inherent nature of human emotions themselves. Emotions, especially those with a strong social component such as anger, fear and disgust, are highly contagious. When individuals are exposed to emotionally charged political content, their own emotions are often triggered, regardless of their initial political leanings. This emotional contagion spreads rapidly, particularly in online environments where individuals are constantly exposed to emotionally charged content from like-minded individuals, reinforcing their pre-existing beliefs and intensifying their emotional responses. This phenomenon is exacerbated by the inherent biases in human cognition, our tendency to seek out information that confirms our beliefs while dismissing information that contradicts them - a cognitive bias known as confirmation bias.

Social media algorithms, designed to maximise user engagement and retention, play a significant role in amplifying this emotional contagion. These algorithms prioritise content that elicits strong emotional responses, whether positive or negative. This means that emotionally charged political content, often characterised by outrage, fear-mongering or ridicule, is prioritised and widely disseminated, creating a feedback loop that reinforces emotional polarisation. Users are presented with a constant stream of information that confirms their biases, creating echo chambers where dissenting opinions are rarely encountered or dismissed out of hand. This creates an environment where individuals are increasingly isolated from alternative perspectives, leading to an intensification of their emotional responses and a further hardening of their political positions.

The spread of misinformation and disinformation exacerbates the problem. False or misleading information, often emotionally charged, is easily shared and amplified on social media platforms, feeding into existing prejudices and reinforcing emotional divisions. The speed and scale with which misinformation spreads makes it extremely difficult to counter, and its emotional impact often overshadows its factual accuracy. For example, the spread of false narratives about electoral fraud or the dangers of vaccines can evoke strong emotional responses such as anger, fear and mistrust, undermining trust in institutions and increasing political polarisation. The emotional impact of misinformation often lasts long after its factual inaccuracy has been debunked.

In addition, the anonymity and lack of accountability often associated with online interactions contribute to the coarsening of political discourse. Individuals are more likely to express extreme and offensive views online, knowing that they are shielded from the immediate consequences of their words. This lack of social restraint fosters a climate of hostility and intolerance, deepening emotional divisions. The proliferation of online trolling and cyberbullying, often targeting individuals with opposing political views, contributes significantly to this toxic environment, discouraging constructive engagement and promoting further polarisation.

The impact of this emotional polarisation goes far beyond the online sphere. It permeates political institutions, making it increasingly difficult to reach consensus on critical issues. The heightened emotional intensity of political debates often overshadows rational arguments, hindering productive policy-making and compromising the effectiveness of government. The inability to engage in constructive dialogue and compromise leads to political gridlock, which further erodes public confidence in democratic institutions. This has a significant impact on the overall social fabric of the nation, as the increasing emotional distance between individuals from opposing political camps contributes to a decline in social cohesion and an increase in social unrest. The consequences can range from increased political violence to a more general social malaise and decline in social capital.

Addressing the issue of emotional polarisation requires a multi-pronged approach. It starts with promoting media literacy and critical thinking skills. Individuals need to be equipped with the tools to critically evaluate the information they encounter online, to recognise the influence of emotional appeals and to identify sources of misinformation. Educational initiatives that focus on emotional intelligence can also make a significant contribution to reducing emotional polarisation. By understanding their own emotions and those of others, individuals can approach political discussions with greater empathy and a willingness to engage in constructive dialogue.

In addition, social media platforms themselves have a crucial role to play. While algorithms designed to maximise engagement may inadvertently contribute to emotional polarisation, these algorithms can also be redesigned to promote more balanced and nuanced political discourse. Implementing effective content moderation policies to combat misinformation and abusive behaviour is essential to creating a healthier online environment. However, striking the right balance between freedom of expression and content moderation remains a complex challenge that requires ongoing dialogue between policymakers, technology companies and civil society organisations.

Finally, political leaders and institutions also have a responsibility to foster a more constructive political climate. By setting a tone of respect and civility in their own communications, they can encourage a more reasoned and less emotionally charged political discourse. Focusing on shared values and common goals, rather than emphasising divisions and differences, can help to bridge the emotional divide and build a greater sense of national unity. Promoting empathy and understanding across the political spectrum, while challenging, is critical to reducing emotional polarisation and fostering a more cohesive and resilient society.

The impact of emotions on political participation is complex and nuanced. It's not just a matter of feeling strongly about a political issue; the type of emotion experienced has a significant impact on the likelihood and form of participation. Anger, for example, is often a powerful impetus for action. When individuals feel wronged or unfairly treated by political systems or policies, anger can act as a catalyst for protest, activism and even violent action. The civil rights movement, fuelled by the righteous anger of individuals facing systemic oppression, provides a powerful example of how this emotion can galvanise collective action and lead to significant social change. Similarly, anger over economic inequality or environmental degradation can lead individuals to participate in demonstrations, boycotts and other forms of political engagement aimed at addressing these grievances. But the expression of anger can also be counterproductive, potentially leading to aggression and violence that undermine the broader goals of political movements. The key is to channel anger constructively, to channel it into effective and peaceful forms of political engagement.

Hope, on the other hand, plays a different but equally important role. Hope, a positive emotion associated with belief in a better future, inspires optimism and fosters a sense of efficacy. Individuals who feel hopeful about the potential for political change are more likely to participate in the political process, believing that their actions can make a difference. Hope can motivate individuals to volunteer for political campaigns, donate to causes they believe in, or actively engage in political discussion and debate. However, excessive hope that is not matched by realistic assessments of political possibilities can lead to disillusionment and disengagement if expectations are not met.

Enthusiasm, a more intense form of positive effect, also plays a crucial role in promoting political participation. Enthusiasm fosters a sense of excitement and engagement with political issues and candidates, leading individuals to actively seek opportunities for engagement. This enthusiasm can translate into volunteering time and resources to support political campaigns, mobilising fellow citizens to participate and organising political events. The passionate engagement of young people in climate change activism, fuelled by a sense of urgency and enthusiasm for environmental protection, illustrates the powerful mobilising power of this emotion.

Political campaigns use enthusiasm strategically to increase voter turnout by creating a sense of excitement and shared purpose among their supporters. However, unchecked enthusiasm can lead to simplistic or uncritical support for a particular candidate or cause, potentially overlooking important details or potential downsides.

Fear, while often associated with negative outcomes, can also have a surprisingly complex influence on political participation. Fear of specific outcomes, such as loss of economic security, societal disintegration or even physical harm, can motivate individuals to engage in political action to protect their interests or values. For example, fears of rising crime rates can lead to increased political activism focused on law enforcement issues.

Similarly, fear of climate change can motivate participation in environmental advocacy groups. However, the use of fear tactics by politicians or political groups, while potentially effective in the short term, often backfires in the long term, eroding trust and leading to cynicism. Fear, when used strategically, can become a tool for political mobilisation, but its ethical implications and potential for manipulation require careful consideration.

Guilt and shame, emotions typically associated with personal failings, can unexpectedly drive political engagement. When individuals feel guilty about their inaction in the face of social injustice or environmental degradation, they may actively seek ways to atone for this perceived moral failing by engaging in political activism or donating to related causes. For example, guilt about the impact of climate change on future generations may lead individuals to become more involved in environmental causes. However, using guilt or shame to manipulate individuals into political action is ethically problematic and can be counterproductive. Genuine moral concern, rather than manipulation, is more likely to sustain engagement in the long term.

The relationship between specific emotions and forms of political participation isn't straightforward. Anger may lead to protest, while hope may lead to volunteering. Context, individual personality differences and the nature of the political issue all play a crucial role in shaping this relationship. An individual who is angry about environmental issues might join a protest, donate to an environmental group, or run for office. The choice of action depends on a variety of factors, including the individual's past experiences, beliefs about their ability to make a difference, and available opportunities for participation.

Moreover, the impact of emotions on political participation is not limited to individual actions. Emotions are contagious, influencing collective behaviour and shaping the overall dynamics of political movements. The contagious nature of emotions, especially in online environments, amplifies the impact of individual emotional responses on collective political action. A shared sense of anger or hope can spread rapidly through social networks, energising political movements and mobilising large-scale participation. This contagion effect is particularly relevant in the context of online activism, where emotionally charged content is easily shared and amplified.

A key element in bridging emotional divides is the promotion of empathy - the ability to understand and share the feelings of others. In the context of political disagreement, empathy involves actively trying to understand the perspectives and emotional experiences of people who hold opposing viewpoints. It requires going beyond simply dismissing or denigrating those who disagree, and instead seeking to understand the underlying reasons for their beliefs and feelings. This is not about agreeing with opposing views, but rather recognising the validity of the emotional experiences that drive those views.

Fostering empathy Active listening, a core component of effective communication, involves paying close attention to what others are saying, both verbally and non-verbally, demonstrating genuine interest in their perspectives, and asking clarifying questions to ensure accurate understanding. This contrasts sharply with the adversarial style of political discourse that often dominates public forums and online discussions, where interruptions, dismissals and personal attacks are common. To foster empathy, individuals need to create space for listening without immediate rebuttal or judgement. This requires conscious effort and practice, especially in contexts where strong emotions are involved.

In addition, reframing political disagreements as challenges requiring collaborative solutions rather than zero-sum conflicts can greatly enhance the potential for constructive dialogue. When political debates are framed as win-lose scenarios, emotions run high and the possibility of finding common ground diminishes. However, framing disagreements as shared challenges encourages the search for mutually beneficial outcomes. For example, rather than focusing on divisive issues such as immigration as a matter of strict border control versus open borders, a more empathetic approach might emphasise the common goal of creating a safe and prosperous society, and then explore different approaches to achieving that common goal. This collaborative approach naturally encourages more constructive dialogue and cooperation.

The power of storytelling also plays a crucial role in bridging emotional divides. Narratives, personal anecdotes and case studies can help humanise abstract policy issues and make them more relatable. Rather than engaging in abstract debates about policies, sharing individual stories of people directly affected by these policies can build empathy and understanding. For example, a story about a family struggling with healthcare costs can evoke more empathy than a statistical report on healthcare inequalities. By highlighting the human impact of policy decisions, stories can soften the rigidity of opposing positions and foster a greater appreciation for the diversity of human experience.

Successful initiatives to bridge political divides often use similar methods. Living Room Conversations, a programme designed to facilitate conversations between people with different political views, uses structured dialogue processes that promote empathy and understanding. Participants are guided through specific questions and prompts designed to encourage active listening, shared storytelling and respectful exploration of different perspectives. The programme has been successful in fostering more positive relationships between participants with previously sharply divergent views, demonstrating the effectiveness of structured approaches to political dialogue. Similarly, initiatives such as the Better Angels organisation work to build bridges across the political divide through non-partisan events, educational programmes and community-building activities.

However, bridging emotional divides also requires addressing the systemic factors that contribute to political polarisation. The role of the media, social media algorithms that reinforce echo chambers, and political rhetoric that deliberately exacerbates divisions cannot be ignored. Media literacy initiatives are crucial to empower individuals to critically evaluate sources of information and identify misinformation or biased reporting. This includes learning to distinguish between fact and opinion, recognising the emotional appeals in political messages and understanding the potential biases inherent in different news sources.

In addition, promoting media literacy helps individuals to recognise and resist the effects of echo chambers - online and offline environments where individuals are primarily exposed to information that confirms their existing beliefs. While social media can be a powerful tool for political engagement, its algorithms often exacerbate polarisation by creating filter bubbles that limit exposure to diverse perspectives. Being aware of these biases and actively seeking out alternative viewpoints is crucial to overcoming the limitations of echo chambers and promoting more balanced perspectives.

Political leaders also have a critical role to play in bridging emotional divides. By using inclusive and respectful language, focusing on shared values and avoiding inflammatory rhetoric, leaders can create a more conducive environment for constructive dialogue and compromise. Emphasising collaborative problem-solving over adversarial conflict can significantly change the tone of political discourse. Leaders can also model empathy and active listening, setting an example for their followers and fostering a more respectful and understanding political culture. Leading by example is crucial in cultivating a more civil and constructive political climate.

Ultimately, bridging the emotional divide in politics requires a multifaceted approach. It involves fostering empathy through individual actions such as active listening and the use of storytelling techniques, as well as working to address systemic factors that contribute to polarisation. Media literacy, constructive dialogue approaches and leadership committed to inclusive communication are essential elements in this endeavour. By cultivating a better understanding of the emotional dynamics of political discourse and consciously employing strategies that promote empathy and shared understanding, we can begin to bridge the divides that threaten effective democratic participation and societal cohesion. The challenge requires sustained effort and commitment from individuals, organisations and political leaders alike. But the potential rewards - a more constructive and inclusive political culture - make investing in bridging these emotional divides a vital and urgent priority.

Defining collective responsibility

Defining collective responsibility requires a nuanced understanding that distinguishes it from individual responsibility. While individual responsibility focuses on the actions and consequences of an individual, collective responsibility focuses on the shared responsibility of a group for its actions, whether those actions are the product of deliberate collaboration or the unintended outcome of a shared context. This distinction is crucial because it requires a shift in focus from individual culpability to a broader examination of group dynamics, decision-making processes, and the intricate web of relationships that shape collective behaviour.

Philosophically, the concept of collective responsibility has been debated for centuries. Some argue that true moral responsibility can only be attributed to individuals, emphasising the importance of individual agency and autonomy. From this perspective, holding a group collectively responsible risks overlooking the nuances of individual actions and motivations, potentially leading to unjust consequences. Others, however, argue that certain actions, particularly those undertaken by groups with a common purpose or structure, transcend individual actions and require collective accountability. This view often emphasises the shared responsibility for creating and maintaining a group environment, even if individual members were not directly involved in a particular harmful act.

The challenge is to balance the need to address systemic wrongdoing with the imperative to avoid punishing individuals who may not have been directly at fault. Interpretations of collective responsibility vary widely in different social and organisational contexts. In political organisations, for example, a political party may be held collectively responsible for policy outcomes, regardless of the specific contributions of individual members. In the workplace, a team may be held collectively responsible for the success or failure of a project, emphasising teamwork and shared responsibility. In social movements, collective responsibility often manifests itself as a shared commitment to a common cause and a collective determination to achieve common goals. The same action can be judged differently depending on the context. For example, a spontaneous act of violence by a crowd may be viewed differently from a well-planned campaign of violence by a structured organisation.

The complexity of assigning blame and credit to groups presents unique challenges. Unlike individuals, groups do not have a single consciousness or intent. Assigning blame or credit to a group requires careful consideration of the group's structure, decision-making processes, and the roles and responsibilities of individual members. This process often involves untangling the complex web of influences and determining the extent to which each member contributed to the collective outcome. Moreover, the inherent difficulties in determining individual contributions within a group can lead to disagreements about who should bear the consequences of collective actions, further highlighting the need for clear mechanisms of accountability and fair procedures for resolving disputes.

The historical evolution of collective responsibility, evident in both positive and negative case studies such as the Nuremberg trials and the Challenger space shuttle disaster, illustrates its multifaceted nature. The concept is not static, but evolves within different social, political and organisational frameworks. The ongoing challenge is to develop more effective and equitable mechanisms for managing collective responsibility, balancing the need for individual accountability with the recognition that many actions are inherently collective in nature. This requires constant reassessment of the ethical dimensions of assigning blame and credit to groups, ensuring that the process is fair, transparent and consistent with broader societal values. Only by carefully exploring these complexities can we harness the potential benefits of collective responsibility while mitigating its inherent risks.

This exploration requires not only a philosophical examination, but also a practical understanding of organisational dynamics, legal frameworks and ethical decision-making processes. The following sections explore these aspects in greater depth, providing a broader understanding of collective responsibility and its implications for individuals, groups and society as a whole.

The benefits of shared responsibility

The inherent benefits of shared responsibility go far beyond the simple division of labour. It fosters a synergistic environment in which the collective output exceeds the sum of individual contributions. This enhancement results from a variety of factors, including increased collaboration, greater efficiency and demonstrably better ethical behaviour. When responsibility is shared, individuals are less likely to feel overwhelmed or isolated, resulting in a more engaged and motivated workforce or social group. This shared burden fosters a sense of camaraderie and mutual support, creating a positive feedback loop where individual successes contribute to collective achievements and vice versa.

Consider the field of scientific research, which is inherently collaborative. Groundbreaking discoveries rarely result from the solitary efforts of a single researcher. Instead, they are the product of teams, often spanning multiple institutions and countries, with each member contributing his or her particular expertise and perspective.

The shared responsibility of the movement fostered a sense of collective agency that enabled the group to overcome systemic oppression and achieve significant social change. The movement's organisational structure, while decentralised, relied heavily on shared accountability, with individuals and groups taking responsibility for specific actions and organising efforts within the broader movement strategy. The sense of collective ownership fostered by this shared responsibility was a crucial factor in mobilising and sustaining the movement's momentum over many years.

The benefits go beyond simply achieving goals; shared responsibility also fosters a culture of ethical behaviour. When responsibility is shared across a group, individuals are less likely to engage in unethical behaviour because they are more aware of the potential impact of their actions on the collective. This 'collective conscience' acts as a powerful deterrent to individual transgressions. Consider a company with a strong ethical culture fostered by shared responsibility. Employees are more likely to report unethical practices or potential conflicts of interest if they know that their concerns will be taken seriously and addressed by the team or organisation as a whole. This shared sense of responsibility reduces the likelihood that unethical actions will go unchecked, protects the organisation's reputation and fosters a more trustworthy working environment. Conversely, organisations with a culture of individual blame often experience a higher incidence of unethical behaviour, as individuals may feel less accountable for their actions.

In addition, shared responsibility often leads to greater efficiency and a fairer distribution of workload. Sharing tasks within a group allows each member to focus on their strengths and areas of expertise, leading to increased productivity and quality of work. It also helps prevent burnout and promotes a more sustainable work rhythm.

Establishing clear communication channels and accountability mechanisms is crucial to reaping the benefits of shared responsibility. Without such mechanisms, shared responsibility can easily degenerate into diffusion of responsibility, where no one feels accountable, resulting in a lack of initiative and reduced productivity. Establishing clear roles and responsibilities from the outset, accompanied by open communication channels to facilitate coordination and feedback, helps to avoid this problem. Regular check-ins and progress reports allow for early identification of potential problems and prompt corrective action, preventing minor issues from escalating into major setbacks. This proactive approach promotes teamwork and a shared understanding of progress, ensuring that the group maintains momentum and focus.

In addition, fostering a culture of trust and mutual respect is critical to the effective implementation of shared responsibility.

When team members trust each other's competence and commitment, they are more likely to work together effectively and support each other. This trust enables individuals to delegate tasks with confidence, knowing that others will carry out their duties diligently. Open communication and transparency in decision-making also contribute to this trust. Regular feedback sessions, where team members can openly express their concerns and suggestions, foster a supportive and collaborative environment. This collaborative atmosphere fosters a sense of shared responsibility and accountability, motivating individuals to contribute their best and work together towards a common goal.

However, the successful implementation of shared responsibility requires careful management and consideration. While there are many benefits, there are also challenges. One potential pitfall is the risk of groupthink, where the desire for consensus overrides critical thinking and the expression of dissenting opinions. This can lead to poor decisions and a failure to identify potential risks. To mitigate this risk, organisations need to foster a culture that values different perspectives and encourages open debate. Mechanisms for anonymous feedback or facilitated discussions can help ensure that all viewpoints are considered and that dissenting opinions are not silenced by social pressure.

Another potential challenge is the difficulty of identifying individual contributions within a collective effort. While collective responsibility recognises shared accountability, it is important to distinguish between individual performance and collective success. Performance reviews and feedback mechanisms should address both individual contributions and the overall performance of the group. This approach ensures that individuals are recognised for their specific efforts, while recognising the importance of teamwork. A balanced approach to recognising both individual and collective contributions promotes fairness and encourages continued participation in collaborative efforts.

Finally, effective implementation of shared responsibility requires a thorough understanding of organisational structure and dynamics. Organisational culture, leadership style and communication channels all play a significant role in determining the success of shared responsibility initiatives. A clear understanding of these factors enables organisations to tailor their approaches to their specific needs and contexts. Organisations can also benefit from implementing innovative strategies, such as establishing shared decision-making platforms or using collaborative work management tools.

Types of collective responsibility

A key distinction is between formal and informal groups. Formal groups, such as corporations, government agencies or military units, operate within established structures, often with codified rules and hierarchies that define roles, responsibilities and accountability mechanisms. Collective responsibility in these settings tends to be more structured and legally defined. For example, collective responsibility for environmental compliance is often enshrined in legal statutes and enforced through regulatory oversight. Failure to comply with these regulations can result in significant financial penalties and legal consequences for the organisation as a whole, even if the violation is due to the actions or omissions of a single individual or a small subset of employees. The organisation's board of directors, senior management and even individual employees may face legal and ethical consequences, underscoring the depth of collective responsibility within formal structures. However, the precise allocation of responsibility can be a matter of considerable legal and ethical debate, depending on the specific circumstances and organisational structure. Internal investigations and audits are often conducted to determine responsibility and apportion blame.

In contrast, informal groups - such as social clubs, community organisations or online forums - lack the formalised structures of their formal counterparts. Collective responsibility is less clearly defined, often relying on social norms, shared values and informal agreements. The consequences of collective action are also less formalised, usually relying on social pressure, reputational damage or ostracism rather than legal sanctions. Consider, for example, a neighbourhood watch group with collective responsibility for maintaining safety. If a member's negligence leads to a security breach, the collective could face reputational damage, leading to a loss of confidence and possibly the dissolution of the group. The consequences are less legally prescribed and rely more on informal social mechanisms. Determining individual responsibility in such cases is often difficult and relies on informal methods of establishing accountability, such as group discussion or self-reporting.

The nature of the action also influences the nature of collective responsibility. Consider actions with immediate and tangible consequences, such as a team's performance in a sporting event.

The collective responsibility for victory or defeat is clear and immediately apparent, with individual contributions having a direct impact on the overall result. Contrast this with actions that have long-term and less obvious consequences, such as a government's environmental policies. Here the collective responsibility of government for the long-term environmental impact of its decisions may be less direct and more diffuse, with many actors and factors influencing the final outcome. Establishing clear lines of accountability between different levels of government and dealing with the multifaceted nature of long-term environmental issues is a major challenge. Attributing specific consequences to particular actions or individuals often becomes a complex task requiring detailed investigation and analysis.

Another important factor is the level of individual involvement in collective action. Consider the spectrum of involvement, from active participation to passive bystander. Those who are active participants tend to have a higher level of responsibility than those who merely acquiesce or remain silent. For example, in a company that engages in unethical practices, active participants who knowingly and willingly participate in those practices bear greater responsibility than employees who passively observe without protest. However, the concept of the "moral bystander effect" suggests that even passive bystanders bear some responsibility for failing to intervene. This responsibility may be less than that of active participants, but it still highlights the ethical obligation to challenge unethical actions in a collective context. The precise degree of responsibility of passive bystanders remains a subject of ongoing ethical and legal debate, often shaped by the specific circumstances and societal norms surrounding the situation. The line between active participation and passive complicity is often nuanced and difficult to define definitively.

An analysis of collective responsibility in the context of corporate social responsibility (CSR) initiatives provides instructive examples. Companies often make a public commitment to ethical and sustainable practices, extending their responsibilities beyond profit maximisation. This covers a wide range of actions, from environmental sustainability to fair labour practices, community engagement and philanthropic endeavours. However, the implementation of these initiatives often reveals challenges related to accountability and the sharing of responsibility.

Community-based environmental projects offer a parallel perspective. These initiatives typically involve collaborative efforts among diverse community members to address environmental challenges. Collective responsibility is critical to their success, but the lack of formalised structures can create challenges in allocating responsibility and ensuring accountability. Individual motivations can vary and the definition of collective goals can be fluid, making it difficult to hold individuals accountable for their contributions or lack thereof. Tracking individual contributions and measuring the impact of collective efforts can also be challenging, as environmental outcomes often involve complex interactions and long-term processes.

Collective responsibility in different contexts

Political organisations, especially political parties, provide a compelling case study. The collective responsibility of a party for its policy outcomes is often implicit, but crucial. Voters hold the party accountable for the success or failure of the policies it implements, regardless of which individual member championed a particular initiative. This collective responsibility extends beyond policy implementation to the party's overall behaviour and public image. For example, if a party is involved in a corruption scandal involving several members, the entire party may suffer reputational and electoral damage, even if only a few individuals were directly involved.

This shared accountability creates incentives for internal mechanisms of oversight and ethical behaviour, although the effectiveness of these mechanisms varies widely across different party structures and cultures. The extent to which individual responsibility can be disentangled from collective culpability becomes a crucial issue in such crises, often leading to internal investigations and, in some cases, significant restructuring within the party organisation.

The allocation of responsibility within a political party is further complicated by its inherent hierarchical structure and power dynamics. While the party as a whole can be held accountable for its actions, assigning individual blame can be difficult.

Senior party leaders often bear a disproportionate share of responsibility, as their influence and decision-making power significantly shape the direction of the party. However, lower-ranking members also contribute to the party's overall performance and actions, raising the question of the extent to which they are collectively responsible for the party's overall success or failure. This ambiguity highlights the importance of clearly defined roles, responsibilities and internal accountability mechanisms within political organisations to avoid situations where individual responsibility is obscured by the weight of collective accountability. Furthermore, the legal and ethical frameworks governing political parties often fail to provide sufficient clarity on the allocation of individual and collective responsibility in the event of wrongdoing, thus fueling ongoing debates about accountability within political systems.

The workplace context presents a different but equally complex manifestation of collective responsibility. Corporations in particular are often held collectively responsible for their actions, both legally and ethically. The concept of corporate social responsibility (CSR) underlines the growing expectation that companies should consider the wider societal impact of their activities and extend their responsibilities beyond profit maximisation. This includes environmental sustainability, labour practices, ethical sourcing and community engagement. However, translating this broad responsibility into concrete individual accountability within the organisation is often a significant challenge. While the company as a legal entity bears ultimate responsibility, it can be difficult to determine the specific contributions and culpability of individual managers, employees or departments involved in specific instances of misconduct. This ambiguity can lead to a diffusion of responsibility, hindering effective accountability mechanisms and potentially encouraging unethical behaviour.

The dynamics of collective responsibility in the workplace are significantly influenced by organisational culture and leadership. A culture that values open communication, transparency and ethical decision-making is more likely to foster a strong sense of collective responsibility and facilitate effective accountability mechanisms. Conversely, a culture that tolerates or even encourages unethical behaviour can lead to a diffusion of responsibility, resulting in increased instances of misconduct. Strong leadership is critical to establishing and maintaining a culture of accountability. Leaders must clearly define expectations, establish clear lines of accountability and effectively enforce consequences for unethical behaviour. In addition, effective leadership encourages open communication and whistleblowing so that potential ethical breaches can be identified and resolved before they escalate into significant problems. However, these mechanisms are not always sufficient, as legal and regulatory frameworks need to complement internal mechanisms to ensure robust accountability.

Social movements offer another fascinating lens through which to examine collective responsibility. Often characterised by decentralised structures and informal networks, these movements present unique challenges in defining and enforcing accountability.

Collective responsibility in this context relies heavily on shared values, goals and a collective commitment to the movement's objectives. However, the absence of formal hierarchical structures and clear lines of authority can make it difficult to establish individual or collective accountability. The actions of individual participants may not always reflect the collective goals of the movement, and it may be difficult to hold the movement as a whole accountable for the actions of individual members. In addition, the inherent diversity of beliefs and strategies within social movements can lead to internal disagreements about the legitimacy and appropriateness of particular actions, making it difficult to establish collective accountability and potentially leading to internal fragmentation.

The success of collective accountability within social movements is often linked to the effectiveness of internal communication, the clarity of collective goals and the development of robust internal mechanisms for conflict resolution and accountability. A strong sense of shared identity and purpose can help foster a sense of collective responsibility, while transparent communication and participatory decision-making processes can help ensure that individuals feel a sense of ownership and accountability for the movement's actions. However, the fluid and often spontaneous nature of social movements makes it difficult to establish formal accountability structures, and the consequences of collective action can be difficult to predict or control.

Collective responsibility is therefore often more diffuse and less formally defined than the formalised structures of corporations or political parties. Measuring the success of collective action within such fluid and ever-changing structures can be particularly challenging.

Comparing the application of collective responsibility in these three contexts reveals significant differences. Political parties operate within established legal and institutional frameworks, although the degree of formal accountability mechanisms can vary considerably. Corporations also have defined legal structures, but their collective responsibility is often extended through ethical norms and CSR initiatives. Social movements, on the other hand, typically lack formal structures, making the implementation and enforcement of collective responsibility a much more complex undertaking.

The degree of success in establishing and sustaining collective responsibility in each context varies considerably and is influenced by factors such as organisational culture, leadership, legal frameworks and the nature of the collective action itself.

In sum, understanding collective responsibility requires a nuanced appreciation of the specific context in which it operates. The application of collective responsibility in political organisations, workplaces and social movements is demonstrably complex, shaped by different structural features, power dynamics and established norms. While each context offers unique opportunities to harness the benefits of shared responsibility, understanding and mitigating the risks of diffusion of responsibility and unequal burden-sharing is critical to ensuring equitable and effective collective action. Further research and careful analysis are needed to fully understand and refine the complexities of collective responsibility in different contexts, ultimately contributing to better practices for ethical behaviour and social well-being.

The ethical dimensions of collective responsibility

One of the most significant ethical dilemmas arises from the potential for inequitable distribution of consequences. When a group acts collectively, the consequences of its actions may fall disproportionately on certain individuals, even those who were not actively involved in the decision-making process or subsequent actions.

Another important ethical challenge is that of moral hazard. When individuals perceive that their personal responsibility is diluted within a larger group, there's an increased risk of moral hazard. This refers to the increased likelihood of taking risks or engaging in unethical behaviour when the costs of such actions are shared or distributed among many. For example, in the workplace, if a team is collectively responsible for completing a project, individual team members may be less inclined to put in their fair share of effort, relying on others to make up for their shortcomings. Similarly, in a political context, if a political party is collectively accountable for its actions, individual politicians may be less inclined to act ethically, knowing that the blame for any wrongdoing will be spread across the party. This phenomenon can seriously undermine the overall effectiveness and ethical standards of the group.

Ethical frameworks provide valuable guidance in navigating these complex ethical dilemmas. Utilitarianism, for example, would advocate the action that maximises overall welfare. In the context of collective responsibility, this might mean distributing the consequences in a way that minimises harm to all, even if this means placing a disproportionate burden on certain individuals.

A deontological approach to collective responsibility might prioritise individual rights and responsibilities, emphasising the need for clear accountability for individual actions. While this approach may ensure individual fairness, it may neglect the wider social implications of collective action. Virtue ethics offers a different perspective, focusing on character development and the cultivation of moral virtues such as honesty, integrity and fairness. From a virtue ethics perspective, cultivating a culture of ethical behaviour within organisations and societies is crucial to establishing effective collective responsibility. This approach emphasises the role of leadership in modelling ethical behaviour and setting clear expectations for ethical behaviour.

It also underlines the importance of education and training programmes to cultivate moral virtues in individuals and promote a sense of shared responsibility and accountability. Addressing the ethical complexities of collective responsibility requires a multi-faceted approach. Clear policies and procedures are essential to ensure a fair distribution of benefits and burdens. Transparency and open communication are essential to avoid misunderstandings and build trust. Robust accountability mechanisms, including clear lines of responsibility and effective enforcement of consequences, are essential to avoid moral hazard. However, these mechanisms need to be designed in a fair and equitable way to avoid scapegoating individuals or unfairly burdening innocent parties.

The role of leadership is critical in shaping ethical behaviour within organisations and societies. Leaders have a key role to play in creating a culture of ethical accountability, where individuals feel responsible for their actions and their contribution to the collective good. This includes setting clear expectations for ethical behaviour, providing training and resources to promote ethical behaviour, and consistently enforcing consequences for unethical behaviour. Leaders who model ethical behaviour themselves are more likely to inspire and motivate their followers to adopt similar standards. The absence of strong ethical leadership can contribute to the diffusion of responsibility and the escalation of unethical behaviour within an organisation or society.

In the political sphere, the challenge is equally significant. Governments are often seen as collectively responsible for the well-being of their citizens. However, the distribution of resources and the implementation of policies can create significant ethical dilemmas. For example, the allocation of resources for public services such as health and education can involve difficult choices that disproportionately affect different segments of the population. Fairness and equity become key concerns, as decisions that benefit one group may disadvantage another. Transparent and accountable decision-making processes are essential to ensure that these decisions are made ethically and in the best interests of all citizens. A lack of transparency and accountability can lead to public distrust and political instability.

Finally, the ethical dimensions of collective responsibility require a thoughtful and nuanced approach. It's not just about assigning blame or distributing consequences; it's about creating a culture of ethical accountability, where individuals and groups are empowered to act responsibly and where mechanisms are in place to distribute the consequences fairly. This requires a concerted effort by individuals, organisations and societies to cultivate ethical values, develop robust accountability mechanisms and establish transparent and equitable decision-making and consequence-sharing processes. Only then can the potential benefits of collective responsibility be fully realised, while minimising the risks of moral hazard and inequitable burden sharing. The journey towards this ideal requires continuous reflection, critical evaluation and adaptation of ethical frameworks to the evolving complexities of human interaction and collective action. The ongoing dialogue on collective responsibility underscores its enduring relevance and the continuing need for ethical frameworks to guide its effective and equitable application.

Loss of individual responsibility

One of the most significant consequences of diffuse responsibility is the emergence of groupthink. Groupthink, a term coined by Irving Janis, describes a phenomenon in which the desire for group harmony and conformity overrides critical thinking and independent judgement. In such situations, individuals may suppress dissenting opinions or critical assessments in order to avoid conflict or maintain a perception of group unity. This can lead to poor decision making, as alternative perspectives and potential risks are not adequately considered. The pressure to conform can stifle creativity and innovation, leading to sub-optimal outcomes and potentially disastrous consequences.

Similarly, in organisational settings, diffused responsibility can lead to a lack of personal ownership of tasks and projects. It is also an important factor in political contexts. When government bodies or political parties are collectively responsible for policy outcomes, individual politicians may be less inclined to take personal responsibility for decisions. This can lead to a lack of transparency and accountability, making it difficult to assign responsibility for successes or failures. It can also encourage risk-taking behaviour, knowing that the potential negative consequences will be shared by many. The lack of individual accountability can contribute to cynicism among citizens and undermine public trust in government institutions.

A related issue is the phenomenon of blame. When things go wrong in an environment of collective accountability, the tendency to blame others is often reinforced. Individuals may try to distance themselves from negative outcomes by pointing the finger at colleagues, superiors or even external factors. This behaviour undermines the effectiveness of collective accountability by preventing meaningful learning from mistakes and hindering efforts to improve future performance. It fosters an environment of mistrust and resentment, making it difficult to build strong, collaborative teams or implement effective solutions to challenges.

To mitigate the risks associated with the loss of individual accountability within a framework of collective responsibility, several strategies can be implemented. First, clear roles and responsibilities should be defined in advance to ensure that everyone understands their specific contributions and obligations.

This clarity can reduce ambiguity and prevent the diffusion of responsibility. Second, it is essential to establish robust accountability mechanisms. This could include regular performance reviews, both individual and collective, to assess individual contributions and overall team performance. Transparent evaluation processes ensure that individual efforts are recognised and rewarded, or that shortcomings are addressed.

It is also important to foster a culture of open communication and constructive feedback. Individuals should feel empowered to raise concerns and dissenting opinions without fear of retribution. Creating safe spaces for discussion and debate can encourage critical thinking and prevent groupthink. Regular team meetings, facilitated by skilled leaders, can serve as a platform for voicing concerns, coordinating tasks and ensuring that everyone is on the same page. These open channels of communication help to ensure that individual contributions are recognised and accountability is maintained.

In addition to fostering open communication, strong leadership is essential to maintaining individual accountability within a collective accountability framework. Leaders should model ethical behaviour, encourage open discussion and provide constructive feedback. They should also actively discourage finger-pointing and promote a culture of mutual support and accountability. Leaders can create a framework in which individual contributions are visibly recognised and rewarded, thereby reducing the tendency for individuals to assume that others will pick up the slack. Such leadership fosters a sense of shared purpose and collective responsibility, while maintaining the importance of individual contribution.

Blame shifting

Blame shifting manifests itself in many ways. It can involve outright denial of personal involvement, a tactic often used by individuals hoping to distance themselves completely from any perceived blame. Alternatively, it can take the form of subtle attribution of responsibility to external factors - a flawed system, unforeseen circumstances or the actions of others. The aim of this strategy is to portray the negative outcome as being beyond the individual's control, thus absolving them of blame. In some cases, blame shifting takes a more direct form, explicitly blaming specific individuals or groups for the shortcomings. This can lead to internal conflict and fracture the very unity that the collective responsibility framework seeks to foster.

The dynamics of blame are deeply intertwined with power structures. Individuals in positions of authority often have a greater capacity to deflect blame, using their influence to shift responsibility downwards within the organisational hierarchy.

Conversely, those with less power are often more vulnerable to blame, regardless of their actual contribution to the negative outcome. This disparity highlights a critical ethical dimension: blame-shifting often falls disproportionately on less powerful members of the collective, exacerbating existing inequalities.

Blame-shifting is not limited to corporate settings; it is equally prevalent in political contexts. Political scandals often involve complex patterns of blame-shifting, where responsibility for policy failures or ethical breaches is distributed among different actors.

Government officials may point the finger at bureaucratic processes, external pressures or even opposition parties. This shifting of responsibility undermines public trust and prevents meaningful accountability.

Ambiguity can demotivate high-performing individuals who feel their contributions are undervalued or overlooked, while allowing less productive members to escape accountability.

To mitigate the challenges of assigning blame and praise in collective contexts, it is essential to foster a culture of transparency and accountability. This means establishing clear roles and responsibilities from the outset, and ensuring that everyone understands their specific contributions and obligations. Regular performance reviews, both individual and collective, are essential to assess individual contributions and the overall performance of the group. These reviews should be conducted in a fair and transparent manner, allowing for open discussion and constructive feedback.

Equally important is the establishment of robust reporting mechanisms. Open channels of communication are essential for reporting concerns, identifying potential problems early and fostering a culture of collective problem-solving. This openness discourages the development of a culture of silence, where individuals are reluctant to report errors or raise concerns for fear of being blamed. A transparent reporting system reduces the temptation to cover up mistakes or shift blame, and promotes a culture of continuous learning and improvement.

The role of leadership in fostering a culture of transparency and accountability cannot be overstated. Leaders should model ethical behaviour, encourage open communication and provide constructive feedback. They must actively discourage blame and create an environment where individuals feel comfortable raising concerns and taking responsibility for their actions. This requires leadership that actively listens to dissenting views, accepts constructive criticism and promotes a culture of mutual trust and respect. Leaders must make a conscious effort to ensure that credit and blame are shared fairly, recognising both individual and collective contributions.

It is also essential to establish a strong ethical framework, including a clear code of conduct. This framework should define acceptable and unacceptable behaviour, clarify expectations and outline potential consequences for breaches. Such a code should be developed collaboratively, with input from all stakeholders, and applied consistently to all members of the collective. Regular training can reinforce ethical standards and raise awareness of potential conflicts of interest.

The role of education in preventing violence

Education is a cornerstone in the fight against violence, both as a preventive measure and as a transformative force. Its impact extends far beyond the classroom, shaping individuals' emotional intelligence, conflict resolution skills and empathy, ultimately contributing to a more peaceful society. The critical role of education in violence prevention begins early, building a foundation of understanding and compassion that serves as a bulwark against aggression. By integrating violence prevention curricula into education systems at all levels, from early childhood to higher education, we equip individuals with the tools they need to manage conflict constructively and build more resilient communities.

The importance of teaching conflict resolution skills cannot be overstated. Children and young adults are often ill-equipped to deal with disagreements peacefully, resorting to aggression as a default response. Education programmes that teach effective communication skills, such as active listening, empathy and compromise, empower individuals to resolve conflicts without resorting to violence.

Empathy, the ability to understand and share another person's feelings, is an essential component of violence prevention. When individuals lack empathy, they are less likely to consider the impact of their actions on others, increasing the likelihood of aggressive behaviour. Teaching children to understand and appreciate different perspectives is crucial to developing empathy. Activities that promote perspective-taking, such as storytelling, literary analysis and community service, can foster empathy and increase social awareness. By providing opportunities for students to engage with different cultures, backgrounds and experiences, education fosters a greater appreciation of human diversity and reduces the likelihood of prejudice and discrimination - two important factors that contribute to violence.

Extending the reach of violence prevention education beyond schools is essential. Embedding these principles in communities strengthens the overall impact of these initiatives.

Promoting a culture of peace and non-violence within educational institutions is as important as specific curricula. This requires a change in the overall school climate that promotes respect, empathy and inclusiveness among students and staff alike. Anti-bullying programmes, conflict resolution initiatives and restorative justice practices create supportive and nurturing environments that help reduce risk factors for violence. The school environment acts as a microcosm of society, allowing violence prevention strategies to be tested and refined. This fosters a safer and more harmonious environment for learning and growth, enabling students to thrive both academically and socially. The overall goal is to create a ripple effect, spreading the positive impact of violence prevention education from the classroom to the community and beyond. By emphasising prevention, education plays a vital and multifaceted role in fostering a more peaceful and harmonious society.

The power of social connections

The diversity of human life is woven with threads of connectedness, and the strength of these threads has a significant impact on our vulnerability to violence. Strong social connections act as a protective buffer, shielding individuals from the harsh realities of stress, adversity and the temptation to resort to aggression. These connections are not just a comforting presence; they are a vital resource, a network of support that builds resilience and promotes positive coping mechanisms. People who are embedded in strong social networks have a greater ability to cope with challenges, thereby reducing the likelihood of violent behaviour.

Research consistently shows a link between strong social support systems and lower rates of violence. Studies of diverse populations have revealed a striking pattern: individuals with extensive and supportive social networks are significantly less likely to engage in violent behaviour, either as perpetrators or victims. This is not just a statistical correlation; the underlying mechanisms are deeply rooted in the basic human need for belonging and connection. Feeling accepted, understood and valued within a supportive community fosters a sense of security and belonging that reduces the appeal of aggression as a means of coping with life's difficulties.

The absence of strong social ties, on the other hand, often contributes to increased vulnerability to violence. Social isolation and loneliness are often cited as risk factors for both perpetrating and experiencing violence. Individuals who feel alienated, marginalised or disconnected from their communities are more likely to experience elevated stress levels, diminished emotional regulation and a reduced ability to cope constructively with adversity. This isolation can exacerbate existing vulnerabilities and increase the likelihood of resorting to violence as a means of expressing frustration, seeking attention or asserting dominance.

The lack of positive role models and supportive relationships exacerbates this risk, making individuals more vulnerable to negative influences and harmful behaviours. Building strong community ties is therefore crucial to preventing violence. This involves fostering a sense of belonging and shared responsibility within communities, promoting social inclusion and actively addressing issues of inequality and marginalisation. A cohesive community is one in which individuals feel connected to their neighbours, share common values and work together to address common concerns. This sense of community fosters trust, cooperation and a willingness to intervene when violence threatens to disrupt the social fabric. Strong community ties create an environment where individuals feel seen, heard and valued, reducing the likelihood of resorting to violence as a means of self-expression or conflict resolution.

Several successful community-based interventions have demonstrated the effectiveness of strengthening social ties in preventing violence. One prominent example is the development of neighbourhood watch programmes, which foster a sense of collective responsibility for safety and security. By encouraging residents to communicate and cooperate in monitoring their surroundings, these programmes create a network of vigilance that discourages criminal activity, including acts of violence. Similarly, community-based mentoring programmes have proven effective in reducing youth violence by providing young people with positive role models and supportive relationships. These programmes match at-risk youth with adult mentors who provide guidance, encouragement and support to help them overcome challenges and develop positive coping mechanisms.

In addition, initiatives that focus on promoting social inclusion and tackling inequalities have shown great promise in preventing violence. By creating opportunities for people from marginalised groups to participate fully in community life, these initiatives reduce social isolation, promote a sense of belonging and reduce the risk of violence. This may involve creating accessible community centres, offering language classes or implementing vocational training programmes targeted at disadvantaged groups. These interventions address the root causes of violence by promoting social justice and creating opportunities for social mobility. The creation of safe spaces, free from harassment and discrimination, is also essential to promote a sense of security and reduce the risk of violence.

The role of social programmes in strengthening social support systems is also crucial. Government-funded initiatives such as after-school programmes, community centres and mental health services can provide essential resources and support networks for individuals and families at risk of violence. These programmes offer a range of services, including youth development activities, parenting education and mental health counselling, which together address multiple risk factors for violence. In particular, access to affordable mental health care is critical to mitigating the effects of stress, trauma and mental illness, which can significantly increase the likelihood of violent behaviour. By addressing underlying mental health problems, social programmes provide a critical layer of prevention, minimising the risk of violence and improving the overall well-being of individuals and communities.

However, the effectiveness of these community-based interventions and social programmes depends on several factors. Strong leadership, community involvement and adequate funding are essential for successful implementation. Sustained commitment and collaboration between different stakeholders, including government agencies, community organisations and residents, is crucial for long-term impact. Evaluation and continuous improvement are also essential aspects of any successful violence prevention programme. Regular assessment of programme effectiveness allows for adjustments and adaptations to ensure that interventions continue to meet the evolving needs of the community.

In addition, the digital age has introduced new complexities into the landscape of social connections. While social media platforms can foster a sense of community and provide access to information and support, they can also contribute to social isolation, cyberbullying and the spread of misinformation, all of which can indirectly increase the risk of violence. Developing strategies for the responsible use of social media, promoting online safety and mitigating the negative effects of online interactions are essential to address the challenges of the digital age. This includes providing education on responsible online behaviour, promoting critical thinking skills to evaluate information, and developing strategies to address cyber bullying and online harassment.

Building supportive relationships

Building strong, supportive relationships is not a passive process; it requires conscious effort, consistent nurturing and a willingness to invest in the connections that enrich our lives. The cornerstone of any healthy relationship, whether family, platonic or romantic, is a foundation built on trust, respect and empathy. Trust, the bedrock of any meaningful connection, allows for vulnerability, honesty and open communication - crucial elements for navigating life's inevitable challenges together.

Respect, recognising the inherent worth and dignity of each individual, fosters understanding and appreciation, preventing the erosion of the relationship through disrespect or disregard. Empathy, the ability to understand and share the feelings of another, creates a space for compassion and support, allowing individuals to feel seen, heard and understood.

Cultivating these core elements requires deliberate practice. Active listening, a skill often overlooked in the frenetic pace of modern life, is paramount. It involves paying full attention to the speaker, not only hearing their words but also observing their body language and emotional cues. Asking clarifying questions and reflecting on what the speaker has said demonstrates genuine engagement and a commitment to understanding their perspective. This practice not only deepens understanding but also creates a safe space for vulnerability and honesty. Avoiding interruptions and giving the speaker sufficient time to express themselves without judgement are equally important aspects of active listening.

Effective communication goes beyond simply exchanging information; it's about expressing needs, feelings and concerns clearly and respectfully. Learning to recognise and articulate personal emotions is a crucial skill in promoting healthy communication. Emotional intelligence, the ability to understand and manage one's own emotions and those of others, is an important predictor of successful relationships. This includes recognising emotional triggers, managing impulsive reactions and developing strategies for navigating difficult conversations without resorting to aggression or defensiveness.

Conflict is inevitable in any close relationship, but how we manage disagreements shapes the trajectory of the relationship. Constructive conflict resolution involves approaching disagreements as opportunities for growth and understanding, rather than as battles to be won. This approach values collaboration and compromise over competition. Active listening remains critical in this context, allowing each party to fully express their perspective before attempting to find solutions. Identifying common ground, even in the midst of differing viewpoints, is key to achieving mutually acceptable outcomes. Compromise, a willingness to meet halfway, is essential in forging solutions that recognise the needs and concerns of all involved. Avoiding personal attacks, name-calling or escalating the conflict is crucial to maintaining respect and preventing irreparable damage to the relationship.

Valuable techniques such as active listening, assertive communication and conflict resolution strategies go beyond the immediate application; these skills are transferable to all aspects of life, improving interactions in personal, professional and social settings.

Learning from successful relationships also provides invaluable insights. Observing healthy relationships in action, whether through personal observation, literature or case studies, offers valuable lessons. Identifying the communication patterns, conflict resolution strategies and support systems used in these successful relationships can serve as a guide for developing one's own healthy relationships. Analysing how individuals in successful relationships manage disagreements, express emotions and provide support reveals patterns of effective communication and conflict resolution. These observations can help individuals identify strategies that match their personal style and communication preferences.

However, it is important to recognise that not all relationships are equally healthy and that some may involve abuse or unhealthy dynamics. Recognising the signs of an unhealthy relationship - controlling behaviour, emotional manipulation, verbal abuse or physical violence - is crucial. Seeking professional help is not a sign of weakness, but of prioritising your own wellbeing.

The rewards of maintaining healthy relationships far outweigh the effort, fostering a sense of belonging, improving overall well-being and contributing significantly to the reduction of violence by providing a strong buffer against the negative effects of stress, isolation and adversity. By prioritising healthy relationships and seeking help when needed, individuals can cultivate a supportive network that acts as a powerful force in preventing violence.

Addressing social isolation and exclusion

Social isolation and exclusion are significant risk factors for violence that are often overlooked in discussions about crime and aggression. A deep sense of disconnection from society, a lack of belonging and the experience of marginalisation can fuel frustration, resentment and a sense of powerlessness. These negative emotions, if left unaddressed, can manifest as aggression and violence, creating a vicious cycle in which isolation breeds further isolation and exacerbates existing social problems. The absence of strong social ties and supportive relationships leaves individuals vulnerable to negative influences and susceptible to harmful behaviours.

Conversely, strong social networks provide a buffer against stress, foster resilience and promote pro-social behaviour. The relationship between social exclusion and violence is complex and multifaceted. People who feel alienated and marginalised are more likely to experience feelings of anger, resentment and frustration, leading to an increased risk of aggressive outbursts. This is particularly true for people from marginalised groups, such as immigrants, refugees, ethnic minorities and people with disabilities, who often face systemic discrimination and social barriers. These barriers can range from difficulties in accessing education and employment to experiences of prejudice and harassment. Such experiences can contribute to feelings of powerlessness, hopelessness and devaluation, increasing the likelihood of resorting to violence as a means of expressing anger or asserting oneself.

The experience of social isolation can also deprive individuals of positive role models and mentors. As discussed in the previous section, positive relationships act as protective factors against violence by providing support, guidance and a sense of belonging. In the absence of such relationships, individuals may turn to negative peer groups or engage in risky behaviours, increasing their risk of involvement in violence. In addition, social isolation can have a negative impact on mental health, exacerbating pre-existing conditions such as depression, anxiety and PTSD, all of which are associated with increased aggression.

Tackling social isolation and exclusion requires a multi-pronged approach that focuses on creating opportunities for social interaction, fostering a sense of community and supporting marginalised groups. A key aspect is to create inclusive environments in schools and workplaces, spaces where everyone feels valued, respected and supported. This means implementing anti-bullying programmes, promoting diversity and inclusion training for staff and fostering a culture of respect and empathy. Schools should actively seek to create a sense of community where students feel they belong and can find support from teachers, counsellors and peers. Similarly, workplaces should prioritise creating a positive and inclusive working environment where employees feel valued, respected and supported. This could include implementing mentoring programmes, encouraging team-building activities and providing resources for employees to address mental health issues.

Integrating immigrants and refugees into the community is another important aspect of reducing social isolation and promoting social cohesion. This includes providing language training, cultural orientation programmes and access to essential services such as healthcare and housing. Initiatives to connect newcomers with established community members through social events, mentoring programmes and volunteering opportunities are also important. Creating welcoming and inclusive communities where immigrants and refugees feel safe, valued and supported is essential to reduce the risk of social isolation and promote successful integration.

In addition, addressing the root causes of social exclusion is crucial for the long-term prevention of violence. This includes tackling systemic inequalities such as poverty, discrimination and lack of access to education and employment. Policies that promote economic opportunity, social justice and equal access to resources are essential to creating a more inclusive and equitable society.

These broader societal changes address the root causes of social isolation and create conditions conducive to social cohesion and violence prevention. These efforts need to be complemented by targeted interventions for marginalised communities.

The effectiveness of these initiatives in reducing violence and promoting social cohesion can be measured using a variety of indicators. Studies can track changes in crime rates, rates of youth violence and incidents of hate crime. Surveys can assess levels of social trust, community involvement and feelings of safety and belonging. Qualitative research can explore the lived experiences of people who have participated in social inclusion programmes, providing rich insights into their impact. Data analysis can reveal correlations between social inclusion initiatives and changes in violence rates, helping to demonstrate the effectiveness of these programmes and identify areas for improvement. Longitudinal studies that follow individuals over time can provide valuable information about the long-term effects of social inclusion initiatives on their well-being and behaviour.

Finally, addressing social isolation and exclusion is a critical component of a comprehensive violence prevention strategy. By creating inclusive environments, supporting marginalised groups and tackling systemic inequalities, we can promote social cohesion and reduce the risk of violence. This requires a concerted effort by governments, communities, schools and workplaces to work together to build a society where everyone feels they belong, are valued and supported. Investments in social inclusion programmes are investments in a safer and more harmonious future, benefiting not only those directly involved but also the wider community. The long-term benefits of such investments far outweigh the initial costs, contributing to a more peaceful and prosperous society for all. The collective responsibility to tackle social isolation and exclusion underlines the importance of a proactive and comprehensive approach to violence prevention.

Understanding individual violence

Understanding individual violence requires a nuanced approach that goes beyond simplistic explanations and recognises the complex web of factors that contribute to aggressive behaviour. While the distinction between individual violence and other forms of aggression may seem straightforward, the line can often be blurred. A spontaneous outburst of anger in a heated argument is very different from a premeditated assault, and both are different from the systematic violence of a hate crime. This subsection focuses on those acts of violence that are driven by individual motivations and explores the psychological, biological and social underpinnings of such behaviour.

A key aspect is to recognise the psychological underpinnings of violence. This includes exploring the role of mental illness, personality traits and learned behaviours. Mental illnesses such as antisocial personality disorder, schizophrenia and bipolar disorder have been associated with an increased risk of violence, although it's important to stress that the vast majority of people with these conditions are not violent. It is not just the presence of a mental illness that matters, but the interaction between the illness, other contributing factors and the individual's specific circumstances. For example, a person with schizophrenia who experiences an acute psychotic episode may be violent, whereas another person with the same diagnosis who receives consistent treatment and support may not. This highlights the importance of taking into account the individual's specific context, including access to treatment, social support networks and coping mechanisms.

Personality traits also play an important role. Individuals who score high on traits such as impulsivity, aggression and sensation seeking are more likely to engage in violent behaviour. However, it's important to remember that these traits alone are not enough to predict violence; they need to be considered alongside other factors such as environmental influences and learned behaviours. A person with a predisposition to aggression may learn to manage and control their anger through therapy or other interventions, effectively reducing the risk of violent outbursts. Conversely, a person with a relatively calm temperament may resort to violence under extreme duress or provocation. Therefore, focusing solely on personality traits provides an incomplete understanding.

Another critical aspect is learned behaviour. Individuals raised in violent homes or exposed to significant trauma often exhibit heightened aggression. The modelling of violent behaviour observed in their environment, whether from family, peers or the media, can significantly shape their own patterns of behaviour. Children who witness domestic violence, for example, may internalise the idea that violence is an acceptable means of conflict resolution.

of conflict resolution. This learned behaviour may manifest in adulthood as patterns of aggression, intimate partner violence or other acts of violence. Similarly, individuals exposed to pervasive societal violence, either directly or indirectly through media portrayals, may develop a desensitisation to violence that reduces their inhibitions to act aggressively.

The impact of stressors such as trauma or abuse is profound and cannot be overstated. Experiences of childhood abuse, neglect or witnessing violence can significantly alter brain development and emotional regulation, increasing the likelihood of violent behaviour later in life. These traumatic experiences often lead to post-traumatic stress disorder (PTSD), which is associated with an increased risk of aggression. The symptoms of PTSD, such as hyper vigilance, flashbacks and emotional dysregulation, can trigger impulsive and potentially violent responses to perceived threats.

In addition, trauma can lead to the development of coping mechanisms that involve aggression or violence as a means of self-protection or expression of overwhelming emotions.

Analysis of real-life case studies highlights the complex interplay of these factors. Consider, for example, the case of domestic violence. Domestic violence rarely has a single cause. Rather, it is often the result of a combination of factors: the perpetrator's history of trauma leading to anger management problems; a personality characterised by impulsivity and aggression; learned behaviours resulting from witnessing violence in childhood; and substance abuse that exacerbates these underlying problems. Further complicating the picture are stressors within the relationship itself, such as financial difficulties or infidelity, which can push an already volatile situation over the edge. Understanding these complex dynamics is essential to designing effective interventions to prevent and respond to domestic violence. Similarly, examining cases of school shootings or workplace violence reveals complex interactions between mental health, perceived injustice, social isolation and access to weapons of all kinds.

The distinction between individual and community violence is not simply a matter of scale. While individual acts of violence contribute to the overall level of violence within a community, the factors that drive them are different. Individual violence results primarily from the psychological, biological and social characteristics of the individual perpetrator, whereas community violence results from broader societal factors such as social inequality, economic hardship and cultural norms. However, the two are not mutually exclusive; individual violence can escalate into community violence, as in the case of riots or gang activity initiated by individual acts of aggression. Understanding this crucial distinction is the basis for developing targeted prevention and intervention strategies.

A closer look at individual violence reveals the limitations of simplistic causal explanations. Attributing violence solely to mental illness, poverty or upbringing ignores the complexity of human behaviour. Rather, it is the interaction of multiple factors - biological predispositions, psychological vulnerabilities, social influences and environmental stressors - that create a high-risk environment for violence. The challenge is not only to identify these factors, but also to understand their complex interactions and to develop interventions that effectively address them.

The role of substance misuse in violent behaviour

The complex relationship between substance misuse and violent behaviour is an important area of research in criminology and social psychology. While substance abuse doesn't cause violence in every case, its role as an important contributing factor, often acting as a catalyst for pre-existing aggressive tendencies or creating situations ripe for conflict, cannot be ignored.

To understand this complex interplay, it is necessary to examine both the pharmacological effects of different substances and the social context in which they are used.

The pharmacological effects of substances such as alcohol and various illicit drugs are well documented and directly relevant to violent behaviour. Alcohol, for example, is a central nervous system depressant that, while initially producing feelings of relaxation and euphoria, can significantly impair judgement, reduce inhibitions and increase impulsivity. This effect is amplified by the intoxicating properties of alcohol, leading to disinhibition and a lowered threshold for aggressive responses to perceived provocation or frustration. Studies consistently show a strong correlation between alcohol consumption and violent crime, with intoxication being a factor in a significant percentage of assaults, domestic violence incidents and homicides. The mechanism isn't simply a matter of 'drunkenness'; rather, alcohol interferes with cognitive functions that govern self-control and emotional regulation, making individuals more likely to act on aggressive impulses without fully considering the consequences.

In addition to alcohol, several illicit drugs have a similar, though often more pronounced, influence on violent behaviour. Stimulants such as cocaine and methamphetamine can cause hyperactivity, paranoia and aggression by increasing dopamine levels in the brain. Users experiencing stimulant-induced psychosis may perceive threats where none exist, leading to violent outbursts in self-defence or in response to imagined aggression. Similarly, hallucinogens such as LSD or PCP can distort perception and induce states of extreme anxiety and fear, making individuals vulnerable to violent reactions born out of a distorted reality. Opiates, while not necessarily directly increasing aggression, can lead to violent behaviour because of their association with illegal drug markets and the desperation often felt by individuals struggling with addiction. The desperate need for the next dose can lead individuals to commit violent acts to secure their supply.

The social context of drug use further complicates the relationship between drug abuse and violence. Drug dealing and gang activity are often inextricably linked, leading to turf wars, retaliatory violence and a general escalation of aggression within communities heavily affected by drug use. The competitive nature of the illicit drug trade, coupled with the potential for high financial rewards and the inherent risks associated with such activities, naturally fosters a culture of violence. In addition, social environments in which drug use is prevalent often lack social cohesion and adequate resources, which can exacerbate existing tensions and provide fertile ground for violent conflict.

Environments characterised by poverty, unemployment and lack of opportunity can exacerbate the effects of drug abuse, creating a vicious circle in which drug use contributes to social dysfunction, which in turn fuels further drug abuse and violence.

Examining specific case studies provides a deeper understanding of this complex relationship. Consider, for example, the many cases of domestic violence in which the perpetrator's alcohol consumption played a key role. While the underlying issues of control, power imbalance and possible past trauma are crucial to consider, the addition of alcohol often acts as an immediate trigger that escalates the situation into a violent confrontation. The impaired judgement and reduced inhibitions caused by alcohol can turn a verbal argument into a physical assault, with tragic consequences. Similarly, substance abuse is a contributing factor in many cases of gang violence. The use of stimulants, for example, can embolden gang members and increase their aggression, leading to escalated confrontations and retaliatory attacks. The combination of social factors such as gang loyalty, territorial disputes and readily available drugs creates a highly volatile environment in which violence is almost inevitable.

Moreover, the relationship between drug abuse and violence is not simply a one-way street. Experiencing violence, whether as a victim or witness, can contribute to the development of drug abuse as a coping mechanism. Traumatic experiences can lead to the development of PTSD, depression and anxiety, leading individuals to self-medicate with substances as a means of escaping the emotional pain and distressing symptoms associated with these conditions. However, this self-medication often exacerbates the problem, creating a feedback loop in which substance abuse contributes to further vulnerability and an increased risk of future violence. The cycle can be particularly pernicious in communities affected by high rates of violence, poverty and lack of access to mental health services.

Understanding the role of substance abuse in violent behaviour requires a multi-pronged approach. Interventions must address not only the immediate problem of substance abuse, but also the underlying social, economic and psychological factors that contribute to both substance dependence and violent tendencies. Effective prevention strategies require comprehensive approaches that focus on early intervention, substance abuse treatment, mental health services and community development initiatives that address issues such as poverty, unemployment and lack of access to health care and education. Addressing the social determinants of health and violence is as important as treating substance abuse itself. By understanding the complex interplay of these factors, we can develop more effective strategies to prevent violence and break the cycle of substance abuse and aggression. A holistic approach, including pharmacological interventions, psychosocial support and community change, is essential to mitigate the significant impact of substance abuse on violent behaviour.

Anger management and impulse control

The pathways to violence are complex, and it is equally important to understand the contribution of deficits in anger management and impulse control. These deficits, often rooted in neurobiological vulnerabilities and exacerbated by adverse life experiences, represent a critical area of focus in the prevention and mitigation of individual acts of violence.

At the heart of aggressive behaviour lies a complex interplay of neurological and psychological processes. As noted above, the amygdala, a brain region associated with the processing of emotions, particularly fear and aggression, plays a central role. In individuals prone to violence, the amygdala may show heightened reactivity to perceived threats, triggering a cascade of physiological responses that prepare the body for 'fight or flight'. This heightened amygdala response can lead to impulsive and disproportionate responses to even minor provocations. At the same time, the prefrontal cortex, which is responsible for executive functions such as planning, decision-making and impulse control, may be less effective at regulating these emotional responses. This imbalance between the heightened reactivity of the amygdala and the weakened regulatory capacity of the prefrontal cortex may result in a reduced ability to control aggressive impulses.

Genetic factors also contribute to individual differences in aggression and impulsivity. While there is no single 'violence gene', studies have identified specific genes associated with variations in neurotransmitter systems, such as serotonin and dopamine, that influence emotional regulation and impulsivity. Variations in these genes may increase an individual's predisposition to aggressive behaviour, making them more vulnerable to triggering events and less able to control their impulses. However, it's important to stress that genetic predisposition is only one piece of the puzzle; environmental factors significantly modulate the expression of these genes.

Adverse childhood experiences (ACEs), such as physical or emotional abuse, neglect and witnessing domestic violence, have a significant impact on the development of emotional regulation and impulse control. Exposure to violence in childhood can disrupt the normal development of the brain's emotional processing centres, leading to increased reactivity to stress and a greater propensity for aggression in adulthood. Children who experience ACEs are more likely to develop mental health disorders, including antisocial personality disorder, which is strongly associated with violent behaviour. In addition, trauma-related stress can alter the neurobiological pathways involved in emotional regulation, making it more difficult for individuals to manage their anger and control their impulses.

In addition to neurobiological and genetic factors, cognitive processes play a crucial role in determining how individuals respond to frustrating or threatening situations. People with poor anger management skills often exhibit cognitive distortions such as catastrophizing (exaggerating the significance of minor events), personalising (taking things personally that are not intended as personal attacks) and all-or-nothing thinking (viewing situations in extreme terms). These distortions fuel negative emotions and increase the likelihood of aggressive behaviour. For example, a person who tends to catastrophize may interpret a minor disagreement as a major personal attack, triggering a disproportionately angry response. Their cognitive distortions hinder their ability to accurately assess the situation and respond appropriately. In addition, limited problem-solving skills may contribute to impulsive aggression. Individuals who struggle to develop effective strategies for coping with challenging situations may resort to violence as a quick, albeit ineffective, solution.

Addressing anger management and impulse control deficits requires a multifaceted approach involving a variety of therapeutic interventions. Cognitive behavioural therapy (CBT) is a highly effective, evidence-based approach that focuses on modifying both cognitive distortions and behavioural patterns associated with aggressive behaviour. CBT helps individuals to identify and challenge their maladaptive thoughts and beliefs, and to replace them with more realistic and helpful perspectives. It also involves practising behavioural techniques, such as relaxation exercises and anger management strategies, to cope more effectively with frustrating situations. Anger management programmes teach people to recognise the early warning signs of anger, to develop coping mechanisms to manage their anger constructively, and to develop alternative strategies for resolving conflict without resorting to aggression.

The effectiveness of these interventions is often demonstrated by a reduction in aggressive behaviour and improved emotional regulation. For example, studies have shown that CBT and anger management programmes can significantly reduce reoffending rates among people with a history of violence. The success of these interventions highlights the importance of addressing cognitive and behavioural factors that contribute to aggressive behaviour. However, the long-term effectiveness of such programmes requires ongoing support and maintenance strategies. Relapse prevention techniques are often incorporated into these interventions to provide individuals with the skills they need to manage their anger and control their impulses in the long term.

The use of these therapeutic approaches is not limited to people with diagnosed disorders. They are also highly beneficial for people who are at risk of violence or who display aggressive tendencies. Proactive interventions that focus on building emotional regulation skills and promoting healthy coping strategies can be highly preventive. These interventions should be tailored to individual needs and circumstances. For example, people with a history of trauma may need trauma-informed therapy that addresses the underlying trauma before directly addressing anger management. Similarly, people with co-occurring substance misuse disorders need integrated treatment that addresses both substance use and anger management issues at the same time.

Reducing social inequalities, improving access to mental health services and promoting community interventions can create a more supportive context in which individuals can effectively manage their anger and control their impulses. A comprehensive approach that integrates interventions at the individual level with changes at the societal level offers the greatest potential for reducing violence. Ultimately, fostering a culture of empathy, conflict resolution skills and access to resources is critical to building safer and more peaceful communities. The complex interplay between individual vulnerabilities and societal influences underscores the need for a multifaceted approach to addressing violence. A combination of effective therapeutic interventions, prevention programmes and societal change is the most promising way to create a society in which violence is significantly reduced and individuals have the resources and support they need to manage their emotions and lead peaceful lives.

Distinguishing between individual and collective violence

Violence is not just an individual phenomenon; it often manifests itself as a collective action shaped by community dynamics and social structures. Understanding this crucial distinction between individual and collective violence is essential to developing effective prevention and intervention strategies. While individual acts of violence certainly contribute to a climate of fear and insecurity within a community, the concept of 'community violence' encompasses a broader range of phenomena that cannot be reduced to the sum of individual acts.

A key difference lies in the context of the violence. Individual violence often stems from interpersonal conflicts fuelled by personal grievances, emotional dysregulation or specific triggers. The causes of individual and community violence are also very different. While individual violence may stem from psychological factors such as anger management problems, trauma or substance abuse, community violence is often rooted in systemic inequalities, social disorganisation and lack of opportunity.

Poverty, unemployment, inadequate education and racial discrimination can all contribute to a climate in which violence is normalised and even expected. Moreover, the absence of strong social institutions such as schools, community centres and law enforcement agencies that foster trust and cooperation can exacerbate this effect, creating "voids" that are often filled by criminal enterprises.

The consequences of these two forms of violence are also different. Individual acts of violence often have an immediate, localised impact on the victim and his or her immediate environment. A robbery, for example, results in direct harm to the victim and may also affect the victim's family and friends. However, community violence can have far-reaching and long-term consequences that extend far beyond the immediate victims. The fear and insecurity created by persistent violence affects social cohesion, economic development and public health. Children growing up in high violence neighbourhoods experience increased trauma, lower educational attainment and an increased risk of becoming involved in criminal activity themselves. This creates a vicious cycle that perpetuates violence across generations.

Furthermore, different types of violence require different prevention and intervention strategies. Addressing individual violence often means focusing on individual-level interventions, such as anger management programmes, substance abuse treatment and mental health services. While these individual interventions are critical, they are not sufficient to address community violence.

Addressing community violence requires a multi-pronged approach that focuses on systemic change and community development. This approach requires addressing the underlying social, economic and political factors that contribute to violence. Initiatives that focus on improving access to education, job training, affordable housing and community-based violence prevention programmes are essential to breaking the cycle of violence at the community level.

at the community level. Involving community members in the design and implementation of these programmes ensures their relevance and effectiveness. It is important to recognise the interconnectedness of individual and community violence. Individual acts of aggression can contribute to a climate of violence within a community, fostering a sense of insecurity and instability. Conversely, community factors can increase the likelihood of individual acts of violence by creating an environment that normalises or even encourages aggressive behaviour. For example, the presence of gangs provides opportunities for individual involvement in violent activities, while the prevalence of firearms can escalate the severity of interpersonal conflicts. Recognising these links underlines the need for a comprehensive strategy that addresses both individual and community level factors simultaneously.

One approach to better understanding these links is to examine the specific pathways through which individual actions contribute to broader patterns of community violence. For example, research could explore the role of 'spill-over effects', where violence in one context (such as domestic violence) spills over into other contexts (such as community violence). Similarly, the concept of 'social contagion' explores how violent behaviour can spread through social networks, creating a cascading effect where individual actions reinforce wider patterns of violence. Understanding these pathways allows for targeted interventions to interrupt the flow of violence from the individual to the community level.

Another critical aspect is to examine the role of institutions and social structures in shaping both individual and community violence. The justice system, for example, can either perpetuate or mitigate violence. Harsh sentencing practices that fail to address underlying social problems can contribute to increased post-release violence, while restorative justice approaches that focus on rehabilitation and community reconciliation can potentially reduce recidivism and promote social healing. Similarly, educational institutions can play a key role in promoting positive social norms and providing opportunities for personal growth and development, thereby reducing the likelihood of individual and community violence.

The media also play an important role in shaping perceptions of violence and influencing behaviour. While the role of the media in directly causing violence is debatable, its portrayal of violence can normalise or sensationalise violent behaviour, contributing to a climate in which violence is perceived as more acceptable or even glamorous. Media portrayals of violence can also influence individual perceptions of risk, creating a sense of fear and insecurity that contributes to a cycle of violence at the community level. Therefore, understanding media representations of violence and their influence on individual and collective behaviour is essential for effective prevention strategies.

Social Inequality and Community Violence

Poverty, arguably the most pervasive form of social inequality, is inextricably linked to community violence. Deprived communities often lack basic resources such as adequate housing, health care, education and employment opportunities. This lack creates a climate of despair and frustration, fostering a sense of hopelessness and resentment that can easily escalate into violence.

In particular, the lack of economic opportunity fuels a cycle of crime and violence. Individuals facing unemployment and underemployment may turn to illegal activities, such as drug dealing or gang membership, as a means of survival or to gain status and recognition within their community. This economic desperation can manifest itself in various forms of violence, ranging from petty theft and assault to more serious crimes such as armed robbery and murder. The statistical evidence overwhelmingly supports this link. Numerous studies have shown a strong positive correlation between poverty rates and violent crime rates in urban areas around the world.

Poverty limits access to resources that might otherwise mitigate violent tendencies, such as mental health care, conflict resolution programmes and opportunities for positive social engagement. Beyond the sheer lack of resources, the experience of poverty often fosters a sense of marginalisation and alienation. Individuals living in impoverished communities may feel disconnected from mainstream society, lacking a sense of belonging or hope for a better future. This sense of exclusion can lead to resentment and anger towards society as a whole, potentially fuelling acts of violence against individuals or institutions perceived as symbols of oppression. This is not to say that poverty directly causes violence in all individuals; rather, it creates an environment in which the likelihood of violence increases significantly. It acts as a catalyst, exacerbating existing tensions and grievances.

In addition, the concentration of poverty in certain geographical areas contributes to a phenomenon known as 'social disorganisation'. In these areas, social institutions such as schools, churches and community organisations can be weak or non-existent, leaving a void that is often filled by criminal enterprises.

The absence of strong social networks and community support systems leaves individuals vulnerable to exploitation and manipulation, increasing their likelihood of becoming involved in violent activities. Research on social disorganisation theory highlights the importance of collective efficacy, the shared belief that neighbours can intervene and maintain order in their community. In high-poverty areas, collective efficacy is often low, contributing to higher rates of crime and violence. This lack of social control allows violent behaviour to flourish unchecked.

Another important factor contributing to the link between social inequality and community violence is discrimination, particularly racial and ethnic discrimination. Studies consistently show that marginalised groups, particularly racial and ethnic minorities, are disproportionately affected by both poverty and violence. This is no coincidence; systemic discrimination often perpetuates cycles of poverty and limits opportunities for economic advancement, thereby perpetuating the conditions that foster violence.

Addressing the link between social inequality and community violence requires a multifaceted approach that addresses the root causes of inequality. This includes policies and programmes aimed at reducing poverty, promoting economic opportunity, increasing access to quality education and health care, and eliminating discriminatory practices in all sectors of society. Community-based initiatives that focus on building social capital, strengthening social institutions and providing conflict resolution services can play a critical role in fostering safer and more resilient communities. These initiatives should be implemented in close collaboration with community members, ensuring that they address the specific needs and concerns of affected populations.

Economic deprivation and its impact

The lack of economic opportunities, the prevalence of unemployment and underemployment, and the severe lack of access to basic resources are not only correlated with violence, but are also important causal factors. Understanding this causal pathway is critical to developing effective violence prevention strategies.

Economic hardship creates a potent breeding ground for violence in several ways. First, the sheer stress of financial insecurity can have a profound impact on mental and emotional well-being. The constant worry about meeting basic needs - food, shelter, clothing - can lead to increased anxiety, depression and irritability. These emotional states can lower inhibitions and increase the likelihood of impulsive, aggressive behaviour. People struggling to make ends meet may be more prone to anger, frustration and a reduced capacity for rational decision-making, increasing the risk of conflict escalating into violence.

This increased stress isn't abstract; it manifests itself in tangible ways. Consider a family struggling to pay rent, facing eviction, and having their children pulled out of school because they cannot afford school supplies and transportation. The despair and anger resulting from such circumstances can easily spill over into aggressive interactions with family members, neighbours or even the authorities. Such situations are not uncommon in communities facing widespread economic hardship. The resulting social friction can easily lead to violent conflict.

Unemployment and underemployment are particularly damaging: the loss of a job or the inability to secure stable employment is more than a loss of income; it's a loss of identity, purpose and social standing. The lack of a regular income can lead to feelings of worthlessness and despair, contributing to mental health problems and increasing the likelihood of violent behaviour. In addition, unemployment often leads to a reduced sense of social inclusion. Employment provides not only financial security, but also social connections and a sense of belonging. Its absence leaves individuals isolated and vulnerable, making them more susceptible to the influence of negative peer groups, including gangs and criminal networks, which can provide a sense of belonging, albeit through illegal activities.

In addition, the lack of economic opportunities creates a climate of hopelessness, especially among young people. When prospects for a better future seem bleak, the lure of quick money through criminal activity can become increasingly attractive. This is particularly true in communities where legitimate employment opportunities are scarce.

Gang involvement often offers a sense of community, status and belonging, providing an alternative social structure in the absence of more positive avenues for social integration. This makes the cycle of violence extremely difficult to break. Lack of economic opportunity fuels the very gangs that perpetuate violence.

Lack of access to basic resources - quality health care, education and adequate housing - further exacerbates the situation. Poor health can lead to increased irritability and aggression, while limited access to education limits opportunities for upward mobility and perpetuates cycles of poverty. Inadequate housing increases stress and insecurity, contributing to a climate of fear and instability. These resource deficits have a cumulative effect, increasing the likelihood of violent behaviour within communities.

The relationship between economic deprivation and community violence is not simply a matter of correlation; extensive research shows a clear causal link. Numerous studies have shown a strong positive correlation between poverty rates and violent crime rates in urban and rural areas worldwide.

Moreover, economic inequality contributes significantly to community violence over and above poverty itself. The stark contrast between wealth and poverty within a community breeds resentment and frustration, creating a sense of injustice and unfairness. This disparity can fuel social unrest, potentially leading to riots, protests and other forms of collective violence. The perceived lack of equality of opportunity exacerbates the sense of marginalisation and alienation felt by people in economically disadvantaged communities.

The impact of economic hardship goes beyond direct acts of violence. The strain on families and communities caused by financial insecurity can lead to the breakdown of social support systems, which are crucial in mitigating conflict. When families struggle to meet their basic needs, the ability to provide emotional support, guidance and supervision to children diminishes, increasing the likelihood of youth involvement in delinquency and violent crime. The erosion of community cohesion weakens social controls, allowing violent behaviour to spread.

Addressing the link between economic hardship and community violence requires a multifaceted approach that addresses the root causes of economic inequality. This includes implementing policies that promote economic opportunity, create jobs and raise the minimum wage to a living wage. Investment in education and training programmes is essential to equip individuals with the skills and knowledge needed to secure stable employment.

The provision of affordable housing and access to quality health care is essential to reduce the stress and insecurity associated with poverty. In addition, community-based initiatives that promote social cohesion and provide support systems can play a critical role in preventing violence. These initiatives could include after-school programmes, mentoring programmes and community centres that provide safe spaces for young people and opportunities for positive engagement. Equally important is investment in mental health services to support those struggling with the emotional toll of economic hardship.

The role of exclusion and discrimination

Marginalisation and discrimination create fertile ground for community violence in a number of ways. First, they create a deep sense of injustice and resentment. When individuals and groups experience systemic barriers to education, employment, housing and health care - barriers rooted in prejudice and bias - deep-seated anger and frustration inevitably arise. This anger isn't just an individual emotional response; it's a collective emotion that reflects the systemic inequalities embedded in the social structure. This collective resentment can easily manifest itself in community-level violence, whether through targeted attacks on members of the dominant group or through broader acts of social unrest and rebellion.

The historical legacy of discrimination plays a significant role in shaping contemporary patterns of violence. Generations of systemic oppression have created deep-seated inequalities in wealth, education and opportunity, making marginalised communities significantly more vulnerable to economic hardship and its associated consequences, as discussed above. These historical injustices continue to manifest themselves in the present through subtle and overt forms of racism, sexism, homophobia and other forms of prejudice, perpetuating cycles of disadvantage and fuelling social unrest. The lasting effects of historical traumas such as slavery and colonialism are manifested in persistent social and economic inequalities, which directly contribute to increased levels of communal violence in affected communities.

Marginalisation also fosters a climate of social exclusion and alienation. When individuals and groups are systematically denied access to mainstream society - excluded from meaningful participation in economic, political and social life - they become increasingly isolated and vulnerable. This isolation can create a sense of powerlessness and despair, increasing the likelihood of resorting to violence as a means of expressing frustration or regaining agency. This sense of exclusion is often compounded by a lack of access to resources such as adequate health care, mental health services and education, further marginalising affected communities.

Lack of social inclusion also contributes to the breakdown of social support systems, making marginalised communities more vulnerable to violence. When trust in law enforcement and other institutions is undermined by experiences of systemic discrimination and biased policing, individuals may be less likely to report crimes or seek help from the authorities, creating a climate of impunity and increasing the likelihood that acts of violence will go unpunished. This mistrust also undermines the effectiveness of community-based violence prevention programmes, making it difficult to build bridges between law enforcement, community leaders and marginalised populations.

The link between marginalisation and community violence is well documented. Numerous studies have examined the disproportionate involvement of marginalised groups in both the perpetration and victimisation of violence. For example, research consistently shows higher rates of homicide, assault and robbery among ethnic minority groups in many countries, often linked to the socio-economic and political marginalisation they face. These disparities aren't just coincidental; they reflect the systemic inequalities that create environments where violence is more likely to occur.

Gang violence in many urban areas is often rooted in the marginalisation and lack of opportunity experienced by young people from disadvantaged communities. Gangs can provide a sense of belonging, identity and protection in the absence of positive social structures and opportunities. This underlines the importance of addressing the root causes of gang violence - marginalisation, economic inequality and lack of opportunity - rather than focusing solely on law enforcement strategies that may not effectively address the underlying social problems.

Religious discrimination can also be a powerful catalyst for community violence. Historical and contemporary examples show how religious persecution and intolerance can fuel intergroup conflict and violence. Marginalisation and persecution of religious minorities can lead to feelings of resentment, insecurity and desire for retribution, resulting in violent clashes between different religious groups or between religious groups and the dominant secular authority. These conflicts highlight the devastating consequences of religious intolerance and the importance of promoting religious tolerance and mutual understanding.

Addressing the link between exclusion and community

Violence requires a multi-pronged approach that addresses the root causes of systemic inequality. This requires a fundamental change in social attitudes and policies, going beyond superficial measures to address the deep-seated prejudices and biases that underpin systems of oppression. Investment in comprehensive anti-discrimination legislation and enforcement is essential, as are proactive measures to promote social inclusion and equal access to resources and opportunities for all marginalised groups.

Equally important is the promotion of intercultural understanding and dialogue. This requires community-based programmes that promote respectful interaction between different groups, challenge prejudices and stereotypes, and create opportunities for people from different backgrounds to connect and build relationships. Promoting media literacy is also essential, encouraging critical examination of media representations of marginalised groups and countering harmful stereotypes that contribute to prejudice and discrimination.

It is also important to address the historical legacy of oppression. This requires acknowledging past injustices, providing reparations where appropriate, and actively working to redress persistent inequalities through policies aimed at achieving social and economic justice. This includes addressing historical land dispossession, wealth disparities resulting from past discrimination, and unequal access to education and health care.

Environmental factors and community violence

The physical environment itself plays a surprisingly important role in shaping the incidence and nature of community violence. This subsection examines the complex interplay between environmental factors and community violence, exploring how the built environment can either encourage or discourage violent behaviour.

High population density, often a feature of urban areas, is strongly correlated with higher rates of crime, including violent crime. The sheer number of people crammed into a limited space increases the likelihood of social friction, interpersonal conflict and, ultimately, violence. It's not just that more people mean more crime; density itself alters social dynamics. In densely populated areas, anonymity is heightened, fostering a sense of detachment and reducing the likelihood of informal social control - the watchful eye of neighbours who might intervene to prevent conflict. This lack of social cohesion and surveillance creates an environment in which acts of violence are more likely to occur, with fewer witnesses and less immediate intervention.

Constant proximity in densely populated areas can also lead to increased stress and competition for resources, exacerbating existing tensions and contributing to a climate conducive to aggression. Consider, for example, the prevalence of gang violence in crowded, impoverished neighbourhoods. Competition for territory, resources and status in these environments often escalates into violent conflict.

In addition to population density, the physical characteristics of a neighbourhood have a significant impact on its propensity for violence. Lack of green spaces, for example, contributes to a more stressful and hostile environment. Studies have shown a strong correlation between access to green spaces, such as parks and recreational areas, and lower crime rates. Green spaces provide areas for social interaction, recreation and stress reduction. They provide a sense of community, creating opportunities for positive social interaction and fostering a sense of collective responsibility. In contrast, areas lacking these vital amenities tend to be characterised by higher levels of social isolation, increased stress and a lack of positive social outlets, contributing to a climate conducive to violence.

Poorly maintained infrastructure - broken windows, graffiti-covered buildings, litter-strewn streets - also plays a crucial role. This is where the broken windows theory of criminology comes in. This theory suggests that visible signs of disorder and neglect - these 'broken windows' - signal a lack of social control and invite further crime. The presence of vandalism, litter and dilapidated buildings sends a message of impunity, suggesting that no one cares about the area and that criminal behaviour is likely to go unpunished. This perception, however inaccurate, encourages further acts of vandalism, which escalate into more serious crimes, including violence. The theory isn't about the broken windows themselves, but the symbolic meaning they convey: a lack of social order and a disregard for the well-being of the community. A well-maintained environment, on the other hand, signals social cohesion and sends a clear message that criminal behaviour won't be tolerated.

The implications of this theory go far beyond simply fixing broken windows. It underlines the crucial role of environmental design and urban planning in crime prevention. Strategic urban planning can actively shape the social environment and reduce opportunities for crime. This includes improving lighting in public spaces to reduce perceptions of vulnerability, creating clear lines of sight to reduce the possibility of concealing criminal activity, and designing public spaces to encourage social interaction and community engagement. Creating well-lit, well-maintained parks, community centres and public transport hubs can significantly reduce opportunities for crime and improve the overall safety and well-being of the community.

In addition to broken windows theory, several other environmental crime prevention theories offer insights into the relationship between the built environment and community violence. Defensible space theory, for example, emphasises the importance of designing environments that allow residents to better control and monitor their surroundings. This can include designing buildings with clear lines of sight, accessible entrances and features that discourage loitering. Similarly, the Crime Prevention through Environmental Design (CPTED) approach advocates the use of environmental design strategies to prevent crime by influencing offender behaviour and reducing opportunities for criminal activity.

This approach includes a variety of strategies such as improving lighting, landscaping and access control to create environments that are less attractive to criminals.

However, the relationship between environmental factors and community violence is not deterministic. These factors create a context, but they don't determine the outcome. Other factors, such as socio-economic conditions, cultural norms and individual

characteristics, all play a role in shaping the level of violence within a community. The impact of the environment is intertwined with these other factors, creating a complex interplay of influences. For example, a poorly maintained environment can exacerbate existing social inequalities, making marginalised communities more vulnerable to violence. Conversely, even in deprived areas, a well-designed and well-maintained environment can contribute to a sense of community and reduce the likelihood of violence.

Tackling community violence therefore requires a holistic approach that addresses both environmental and social factors. Improving the physical environment alone is not enough; it must be combined with strategies that address the root causes of violence, such as poverty, inequality and lack of opportunity.

Effective interventions must be multi-pronged and include environmental improvements, social programmes, economic development initiatives and law enforcement strategies. In addition, community involvement is critical to the success of any violence prevention initiative. Community-based programmes that empower residents to take an active role in shaping their environment are more likely to be successful in reducing crime and promoting community safety.

Ultimately, the relationship between environmental factors and community violence underlines the importance of a comprehensive approach to crime prevention. It's not just about tackling the symptoms of violence; it's also about creating environments that foster social cohesion, reduce opportunities for crime and promote a sense of community ownership and responsibility. A combination of well-informed urban planning, community engagement and social interventions is crucial to creating safe and thriving communities where violence is minimised. Ignoring the powerful influence of the environment would be a serious oversight in any serious attempt to tackle community violence. A truly effective strategy must address the physical, social and economic dimensions of the problem in a holistic and integrated manner.

The impact of community disorganisation

While the physical characteristics of a neighbourhood undoubtedly contribute to its propensity for violence, the social fabric itself - the strength of its ties, the effectiveness of its leadership and the efficiency of its institutions - is proving equally, if not more, important. Community disorganisation, characterised by weak social ties, a lack of collective efficacy and ineffective governance, creates a fertile ground for crime and violence to flourish. This breakdown in social order facilitates a climate of impunity, where criminal acts are less likely to be deterred or reported, and where offenders are less likely to face consequences.

A key element of community disorganisation is the weakening of social ties. Strong social networks, characterised by trust, mutual support and shared responsibility, provide an important buffer against violence. In cohesive communities, neighbours know each other, look out for each other and are more likely to intervene to prevent or report criminal activity. This informal social control, often referred to as 'collective efficacy', significantly reduces the likelihood of violence. When these ties are weak, a sense of anonymity prevails, fostering a climate in which individuals feel less responsible for their actions and in which criminal behaviour is less likely to be challenged. The absence of strong social ties deprives communities of the crucial mechanisms that typically deter violence. This absence is particularly evident in rapidly changing neighbourhoods where established social networks have been disrupted by population shifts, immigration or economic upheaval.

The lack of effective community leadership exacerbates the problem. Strong leadership provides a framework for collective action that enables communities to address social problems and promote safety. Effective leaders facilitate communication, mobilise resources and coordinate efforts to prevent and respond to crime. They build consensus, foster a sense of common purpose, and advocate for the needs of their community. In contrast, a lack of leadership creates a vacuum, leaving communities vulnerable to the influence of criminal elements and hindering their ability to organise collective responses to violence. This lack of coordinated effort makes it more difficult to address issues such as gang activity, drug trafficking and other forms of organised crime that often contribute to community violence.

Ineffective policing also plays a significant role in community disorganisation and its contribution to violence. When the police are perceived to be unresponsive, unfair or even hostile towards certain sections of the community, trust erodes and cooperation diminishes. This breakdown in trust hampers the police's ability to effectively prevent and investigate crime, including violence. In such environments, residents are less likely to report crime or cooperate with investigations, leading to a cycle of impunity and escalating violence. Conversely, effective policing, characterised by community engagement, transparency and accountability, fosters trust and improves crime prevention efforts.

Policing strategies that focus on building relationships with community members, addressing underlying social problems and using community policing models are far more effective in reducing violence than strategies that rely solely on punishment and suppression.

In many rapidly growing urban areas around the world, the influx of new residents often overwhelms existing social structures, resulting in a breakdown of community cohesion and an increase in violence. In such cases, the lack of integration and the absence of established social networks make communities more vulnerable to criminal activity.

Tackling community disorganisation requires a multi-faceted approach that addresses its root causes. Strengthening social ties can be achieved through various initiatives such as community centres, neighbourhood watch schemes and faith-based organisations.

These initiatives provide opportunities for residents to interact, build relationships and develop a sense of collective responsibility. Strengthening community leadership is equally important. This can be done by supporting local organisations, providing leadership training, and creating opportunities for residents to participate in decision-making processes. It is imperative to improve police-community relations through community-oriented policing strategies that promote transparency, accountability and build trust between law enforcement agencies and the communities they serve.

In addition, addressing the underlying social and economic factors that contribute to community disorganisation is essential for long-term solutions. This includes initiatives that focus on job creation, affordable housing, education and access to health care. By addressing these factors, we can create a more equitable society in which communities are better equipped to prevent and respond to violence. Ultimately, the success of any strategy to reduce community violence depends on the strength and resilience of the community itself. Empowering communities through engagement, investment and effective governance is central to promoting safe and thriving neighbourhoods.

The link between community disorganisation and violence is undeniable. The absence of strong social bonds, effective leadership and responsive policing creates a climate of impunity in which criminal behaviour thrives. Tackling this complex issue requires a holistic approach that simultaneously addresses social, economic and environmental factors, while focusing on empowering communities to build resilience and foster strong social cohesion. The ultimate goal is to move from reactive responses to violence to proactive community-building strategies that lay the foundations for sustainable peace and security. This means investing in programmes that build social capital, strengthen community institutions and foster collaborative partnerships between residents, law enforcement and local government. Only through this comprehensive approach can we effectively mitigate the effects of community disorganisation and create safer, more vibrant communities for all.

Evidence shows that the effectiveness of such interventions is linked to their tailoring to specific community needs and cultural contexts. A 'one size fits all' approach is unlikely to be successful. Rather, tailored interventions developed in close consultation with community members are essential to ensure the sustainability and effectiveness of these initiatives. This requires community-led assessments of local challenges that promote a bottom-up approach in which residents are actively involved in identifying priorities and shaping the design and implementation of violence prevention strategies. Such participatory processes promote ownership and increase the likelihood of programme success by building on the strengths and resources already present in the community.

Finally, the long-term effectiveness of any community violence intervention requires ongoing evaluation and adaptation.

Regular monitoring of key indicators, such as crime rates, social cohesion measures and levels of community participation, is critical to measuring the impact of interventions and making necessary adjustments. This continuous feedback loop allows for a dynamic and responsive approach, adapting strategies as circumstances change and new challenges emerge. It ensures that resources allocated to violence prevention are used efficiently and effectively, maximising their impact on the safety and well-being of the communities they serve. Ongoing evaluation also informs future policy and programme development, providing an evidence-based framework for reducing community violence in the long term.

Gang dynamics and violent crime

Understanding gang violence requires an appreciation of the complex web of social dynamics, power struggles and individual motivations that fuel these criminal enterprises. Gangs are not simply random collections of individuals; they have distinct structures, hierarchies and codes of conduct that shape their behaviour and contribute to the high levels of violence associated with them. These structures often mirror those of more formal organisations, with clear leadership roles, specialised functions and defined territories.

At the top of the hierarchy is the leadership, often made up of experienced and ruthless individuals who control the gang's operations, make strategic decisions and settle internal disputes - often violently. Below them are various ranks and roles, including enforcers, responsible for maintaining order and carrying out violent acts; recruiters, tasked with attracting new members; and foot soldiers, who carry out a range of tasks from dealing drugs to carrying out attacks on rival gangs. This hierarchical structure creates a sense of order and control within the gang, but it also fosters a climate of violence in which advancement often depends on demonstrating loyalty and a willingness to engage in violent acts.

Recruitment strategies are critical to understanding gang growth and the perpetuation of violence. Gangs actively seek out vulnerable individuals, often those who are marginalised by society and who are susceptible to manipulation and coercion. These individuals may come from disadvantaged backgrounds, lack positive role models, experience feelings of alienation and isolation, or be searching for a sense of belonging and identity. Recruitment techniques can range from subtle manipulation and promises of loyalty, protection and camaraderie to more coercive tactics, including intimidation and threats. Young people are particularly vulnerable to these recruitment strategies, often attracted by the lure of power, status and a perceived sense of community that traditional social structures may not provide. Understanding these recruitment strategies is critical to designing effective prevention programmes that target at-risk youth and provide them with alternative pathways to success and self-esteem.

The concept of "turf" plays a central role in gang violence. A gang's territory represents not only physical space, but also a source of power, economic opportunity and prestige. Control of certain areas allows gangs to engage in illegal activities - such as drug dealing, extortion and theft - with less risk of interference from rival gangs or law enforcement. Territorial disputes therefore often lead to intense and prolonged conflict, with violence escalating as gangs compete to establish and maintain dominance over key locations. These disputes often go beyond simple skirmishes and turn into organised campaigns of intimidation, assault and murder as gangs fight to assert control and defend their economic interests. The violence is not always spontaneous; it can be meticulously planned, involving surveillance, strategic ambushes and retaliatory attacks. This pattern of violence highlights the critical role of territory in shaping gang activity and fuels the perpetual cycle of violence that plagues communities affected by gang presence.

The escalation of gang violence is often fuelled by factors that go well beyond the immediate dynamics of gang rivalry. Socio-economic factors such as poverty, unemployment and lack of opportunity create an environment in which gang membership becomes an attractive, albeit dangerous, alternative for many young people. In such communities, gangs offer a sense of belonging, purpose and material resources that may not be available through legitimate channels. This lack of opportunity is often compounded by systemic inequalities, creating a vicious circle in which marginalised communities are more likely to be affected by gang violence, and gang involvement exacerbates existing social and economic disadvantages. The lack of positive role models, combined with easy access to weapons and exposure to violent behaviour within the community, further contributes to the cycle of violence and creates an environment ripe for recruitment and gang-related crime.

Furthermore, the relationship between gang activity and wider societal factors cannot be overlooked. For example, the prevalence of drug trafficking is often intertwined with gang violence, as gangs compete for control of lucrative drug markets in their territories. This competition escalates conflicts and leads to extreme violence as gangs fight for their share of the profits. Similarly, the availability of firearms plays a crucial role in the lethality of gang violence. Easy access to guns increases the potential for lethal consequences, turning relatively minor disputes into deadly confrontations. This underlines the importance of addressing the societal factors that contribute to gang violence, such as tackling the root causes of poverty and inequality and strengthening gun control laws.

Prevention and intervention strategies must be multi-pronged, addressing both the immediate dynamics of gang violence and the underlying social and economic factors that fuel it. Focusing solely on law enforcement suppression, while a necessary component, is not sufficient to achieve lasting reductions in gang-related violence. Interventions should include community-based programmes that provide alternative opportunities for young people, such as mentoring programmes, vocational training and educational support. Addressing the root causes of gang involvement by offering positive alternatives is essential to breaking the cycle of violence and providing at-risk young people with the tools and support they need to escape the lure of gang life. These initiatives must be tailored to the specific needs and circumstances of each community, involving community stakeholders and building strong partnerships between law enforcement, social service agencies and community leaders.

Strategies aimed at dismantling gang structures are also needed. This involves identifying and targeting gang leaders and key operatives, disrupting their operations and dismantling their networks.

This work requires intelligence gathering, targeted law enforcement and sustained efforts to address the systemic issues that allow gangs to flourish. However, the long-term success of these strategies depends on addressing the root causes of gang violence and creating conditions that make gang membership less attractive to young people. By improving socio-economic conditions, providing alternative opportunities and strengthening community ties, we can build more resilient communities that are less vulnerable to gang recruitment and violence.

Territorial conflicts and their escalation

Territorial disputes are at the root of much gang violence. The concept of 'turf' transcends mere geography and represents a complex interplay of power, resources and prestige. Control of a territory gives a gang access to illicit activities such as drug trafficking and extortion, provides a source of income and consolidates its position within the criminal underworld. This control isn't just passive; it's actively defended, often with brutal force, against encroaching rivals. Competition for these lucrative territories fuels a perpetual cycle of violence in which seemingly minor infractions can escalate into major conflicts, resulting in injury and death.

Escalation is rarely spontaneous. It's often a calculated, multi-step process that begins with subtle incursions and escalates into open warfare. Initial provocations may involve graffiti, the symbolic marking of territory to assert dominance and send a warning to rivals. These acts, though seemingly insignificant, are powerful signals of intent. They represent a challenge to the established order, a test of the established gang's resolve. If left unchallenged, these incursions can embolden the aggressors and lead to more significant acts of aggression.

The response of the established gang will determine whether the conflict remains local or escalates into a full-scale war. A strong, immediate response may deter further provocations, thereby strengthening their control over the territory.

Conversely, a weak or delayed response may encourage the rival gang to push the boundaries further, perhaps initiating more direct forms of confrontation such as assaulting gang members or vandalising their property. This escalating pattern of action and reaction creates a volatile dynamic in which each act of violence provokes a more severe response, spiralling the conflict into a cycle of retaliatory attacks.

Social hierarchies within gangs also play an important role in the escalation of territorial disputes. Younger, more impetuous gang members may be eager to prove their loyalty and gain status within the gang. They may initiate confrontations without consulting the leadership, potentially triggering a retaliatory response from the rival gang that the leadership has to manage. This dynamic can put immense pressure on gang leaders to respond decisively to maintain their credibility and authority, even if this response escalates the conflict.

Competition for resources is another critical element in the escalation of territorial disputes. Gangs often fight over lucrative drug trafficking routes, sites for extortion rackets, or areas with high footfall for other illegal activities. The economic benefits associated with controlling certain areas can be substantial, making the stakes incredibly high. The value of territory translates directly into the intensity of conflict, with richer territory leading to fiercer fighting. The potential for significant financial gain fuels the intensity of rivalry, increasing the likelihood of violence and even murder.

The strategies used by gangs to defend their territories vary widely.

They range from passive strategies such as the strategic placement of lookouts, to active defence mechanisms such as ambushes, coordinated attacks and the use of weapons such as firearms, knives and even improvised explosive devices. They may use intimidation tactics such as threats and public displays of violence to deter rivals. Some gangs invest in surveillance and intelligence networks to monitor rivals' movements and prepare counter-measures. Other gangs cultivate alliances with other criminal groups to strengthen their defences. The choice of strategy often depends on the gang's resources, the nature of the threat and its overall strategic objectives.

The escalation of gang violence in territorial disputes is not a linear process; it is characterised by periods of relative calm punctuated by sudden outbursts of violence. These outbursts of violence often occur in response to specific events, such as an incursion into a gang's territory, the assassination of a key member, or a change in the power dynamics within the gang. Understanding these triggers is crucial to preventing escalation, although they are inherently difficult to predict with certainty.

The consequences of territorial disputes extend far beyond the immediate participants. Innocent bystanders can be caught in the crossfire and suffer injury or death. The fear and insecurity generated by gang violence can disrupt daily life, affecting businesses, schools and community activities. The economic costs of gang violence are substantial, including health care costs, lost productivity and the cost of law enforcement responses. In addition, the pervasive sense of fear can lead to social fragmentation, making it difficult for communities to rebuild and thrive.

Addressing territorial disputes and their escalation into violence requires a multifaceted approach. This includes not only law enforcement efforts, but also comprehensive social programmes aimed at addressing the root causes of gang violence, such as poverty, inequality and lack of opportunity. Community-based initiatives that focus on providing young people with alternative pathways to success, including education, vocational training and mentoring programmes, can play a crucial role in preventing youth involvement in gangs. Strengthening community ties and creating a sense of collective responsibility can help mitigate the factors that contribute to gang recruitment and violence. In addition, stricter gun control measures can help reduce the lethality of gang-related violence.

Gang recruitment and socialisation

The pervasive influence of gangs on communities goes beyond the immediate violence they inflict; it deeply affects the lives of individuals through recruitment and socialisation processes. Understanding these processes is critical to developing effective interventions and breaking the cycle of gang violence. Gang recruitment is rarely a random event; it's a carefully orchestrated process that exploits vulnerabilities and offers appealing, if deceptive, incentives.

A key driver of gang recruitment is the search for belonging and identity. Many young people, particularly those from marginalised communities, feel alienated from mainstream society.

They may lack positive role models, feel disconnected from their schools or families, or struggle with feelings of inadequacy and low self-esteem. Gangs provide a sense of community, a readily available social network and a structure that fills this void. The camaraderie, shared experiences and sense of belonging that gang membership provides can be very attractive to vulnerable young people, especially those who lack positive support systems elsewhere. The rituals, symbols and shared identity foster a strong sense of belonging and cohesion, creating a powerful force that binds members together. This sense of belonging is often more compelling than the risks associated with gang involvement.

The promise of protection is another powerful attraction for potential recruits. In neighbourhoods plagued by violence and poverty, gangs can offer a perceived safety net. Gang membership can provide a sense of security, protecting individuals from victimisation by rival gangs or other threats. This protection can take many forms, from physical defence against attacks to help with everyday challenges such as finding resources or navigating complex social situations. For young people growing up in high-crime areas, the prospect of gang protection may outweigh the risks of legal consequences or involvement in violent activities.

This is particularly true in areas where the presence of formal protection and law enforcement is minimal or perceived as ineffective. The perception of protection offered by gangs can thus be a powerful recruitment tool.

Economic incentives also play an important role in gang recruitment. Gangs are often involved in a range of criminal activities, such as drug dealing, theft and extortion, which provide financial opportunities for their members. For young people from impoverished backgrounds with limited access to legitimate employment, the lure of quick money and material possessions can be incredibly compelling. This economic incentive can be particularly strong in communities with high unemployment and limited access to education and training. Gangs can become a source of income and a means of social mobility, offering a stark contrast to the perceived lack of opportunities within the formal economy. The promise of financial gain, however meagre or risky, can be a powerful incentive, especially for young people who see few alternatives.

Once recruited, gang socialisation begins, shaping an individual's values, beliefs and behaviour. This process involves a complex interplay of factors, including exposure to violence, peer pressure and the reinforcement of gang-specific norms and values. Exposure to violence is almost normalised within the gang environment. Acts of violence are often justified as necessary for survival, self-defence or maintaining gang loyalty. This desensitisation to violence is a critical component of gang socialisation and contributes to the perpetuation of a cycle of violence. Normalising violence through repeated exposure and justifying violent acts as necessary or even honourable reinforces the gang culture of violence.

Peer pressure plays a crucial role in reinforcing gang norms and behaviour. Members are expected to conform to the group's rules and expectations, including the use of violence when necessary. The consequences of non-compliance can range from social ostracism to physical violence. The pressure to conform is reinforced by the strong social bonds within the gang, creating an environment where deviating from the norm is risky and challenging. This strong peer pressure can lead individuals to commit acts of violence that they might not otherwise consider. Fear of rejection or reprisal can override an individual's moral compass and lead them to participate in violent acts.

The reinforcement of gang-specific norms and values is another key aspect of socialisation. Gangs often promote a code of conduct that values loyalty, strength and dominance. This code emphasises the importance of protecting gang members, retaliating against rivals and maintaining control of territory. Adherence to these values is continually reinforced through various rituals, ceremonies and shared experiences. These rituals instil a deep sense of loyalty and commitment to the gang, often overriding any moral constraints or personal objections to violence.

The process of gang socialisation is often gradual and insidious, subtly shaping the individual's worldview. Over time, violent behaviour becomes ingrained and the individual's identity becomes increasingly intertwined with gang membership. The consequences of this process can be devastating, leading to long-term involvement in criminal activity, imprisonment and even death.

Tackling gang recruitment and socialisation requires a multi-faceted approach that combines prevention strategies with intervention programmes. Prevention initiatives should focus on addressing the underlying factors that make young people vulnerable to gang recruitment. This includes providing access to education and vocational training, creating positive role models within the community, and strengthening families and community support networks. Such initiatives must be tailored to the specific needs of at-risk youth and communities.

Addressing strategies to prevent and intervene in gang violence

Tackling the pervasive problem of gang violence requires a multi-pronged approach that goes beyond simple law enforcement strategies. Effective solutions require a deep understanding of the underlying social, economic and psychological factors that drive gang involvement, coupled with comprehensive prevention and intervention programmes tailored to the specific needs of the community. This requires a collaborative effort involving law enforcement, social service providers, educators, community leaders and, most importantly, the individuals and communities directly affected by gang violence.

Community-based programmes are a critical first line of defence against gang recruitment and escalation of violence. These programmes should focus on identifying and addressing the root causes of gang involvement, such as poverty, lack of opportunity and social isolation. Successful programmes often take a holistic approach, providing a range of services that address the multiple needs of at-risk young people and their families.

One successful model is the development of community centres that offer educational enrichment, recreational activities and mentoring programmes. These centres provide a safe and supportive environment for young people, offering alternatives to the lure of gang membership. By providing positive role models and opportunities for personal growth, these centres can divert young people from gang involvement and foster a sense of community pride and belonging.

The effectiveness of these centres is enhanced if they are actively integrated into the fabric of the community, involving local residents in their planning and operation. This fosters a sense of ownership and responsibility and creates a stronger sense of collective efficacy - the belief that community members can work together to solve problems.

Another key element of effective community-based programmes is the provision of training and employment opportunities. For many young people, the appeal of gangs lies in the perceived promise of quick money and material possessions. By providing access to vocational training, apprenticeships and employment services, communities can offer viable alternatives that address the economic incentives that drive gang involvement. This is particularly important in communities with high unemployment and limited access to education and training. The focus should be on skills development that leads to sustainable employment, not just temporary jobs. Programmes that provide ongoing support and mentoring after job placement are more likely to be successful in the long term.

Equally important are programmes that strengthen families and community support networks. Strong families are often a critical buffer against the negative influences of gangs. Family-focused interventions can include parenting skills training, family counselling and support groups. These programmes aim to improve communication, resolve conflicts and strengthen family bonds. Community-based initiatives that promote positive adult-youth relationships, such as mentoring programmes, can also play an important role in building resilience to gang involvement. Mentors can provide guidance, support and encouragement to help young people overcome challenges and develop positive life skills. The success of mentoring programmes depends on carefully matched pairs and ongoing monitoring of the relationship.

Law enforcement initiatives play a critical role in tackling gang violence, but their effectiveness is greatly enhanced when they are integrated into a broader, community-based strategy. Focusing solely on arrests and imprisonment often fails to address the underlying causes of gang violence and can actually exacerbate the problem by alienating communities and reinforcing cycles of violence. A more effective approach is one that emphasises community policing, building trust and relationships between law enforcement and community members. This involves regular interaction with community members, actively listening to their concerns and working together on strategies to tackle gang-related problems. Community policing initiatives should focus on proactive measures, working with community leaders and residents to prevent gang activity before it escalates into violence. This could include addressing issues such as graffiti, drug dealing and other indicators of gang presence in a community.

In addition to community policing, effective law enforcement strategies include targeted interventions against gang leaders and key figures. Disrupting gang leadership can weaken the gang's structure and reduce its capacity for violence. However, such strategies must be carefully planned and implemented to avoid unintended consequences, such as provoking retaliation or increasing gang cohesion. It is essential that law enforcement activities are aligned with the broader community-based approach, building trust and cooperation rather than creating antagonism. Targeted interventions should be focused, using intelligence-led policing to effectively disrupt gang operations. The use of technology and data analysis can play a crucial role in this process.

Societal impacts of violence Erosion of trust and social cohesion

The economic consequences of violence, as detailed above, are a significant part of its overall impact. However, the damage inflicted goes far beyond monetary losses; it has a profound impact on the fabric of society, eroding trust, fracturing social cohesion and undermining the sense of community that is essential for a thriving population. The societal impact of violence is insidious and long-lasting, often manifesting itself in ways that are difficult to quantify, but no less devastating.

One of the most pervasive effects is the erosion of trust. When violence becomes commonplace, a pervasive atmosphere of fear and suspicion takes root. Neighbours become wary of each other, reluctant to engage in the spontaneous interactions that build strong communities. The simple act of walking down the street can become fraught with anxiety as residents constantly scan their surroundings for potential threats. This constant vigilance creates a sense of isolation and alienation, weakening the bonds that bind individuals and neighbourhoods together.

Children growing up in such environments can develop a heightened sense of fear and mistrust, affecting their social development and ability to form healthy relationships. This erosion of trust extends beyond personal relationships to institutions. Confidence in law enforcement, the justice system and even local government can diminish, creating a cycle of cynicism and disengagement that further undermines social cohesion. Individuals may be less likely to report crimes or cooperate with the authorities, hampering law enforcement efforts and perpetuating a climate of impunity.

The breakdown of trust is closely linked to increased social fragmentation. Violent incidents often cause deep divisions within communities, exacerbating existing inequalities and creating new fault lines. Different ethnic or socio-economic groups may become increasingly isolated from one another, reinforcing stereotypes and prejudices. This segregation can lead to a lack of understanding and empathy, fostering resentment and mistrust. In addition, violence can disproportionately affect vulnerable populations, such as marginalised communities or people with mental health problems, widening existing inequalities and creating a sense of injustice. The sense of powerlessness and lack of control often experienced by those living in areas of high violence contributes to social fragmentation and hinders collective efforts to address the problem.

The consequences of violence extend beyond the immediate victims and perpetrators, affecting the entire social fabric. Businesses may be reluctant to invest in violence-affected areas for fear of property damage or disruption to their operations. This can lead to economic stagnation and a decline in community services, exacerbating existing inequalities. Parents may struggle to provide a safe and nurturing environment for their children as the constant threat of violence weighs heavily on their minds. Schools in high-crime areas may find it difficult to provide a quality education as pupils struggle to concentrate amidst a pervasive sense of insecurity. This cascade effect has a negative impact on the overall health and well-being of the community, affecting its social, economic and psychological development. The cycle of violence and its social consequences is self-perpetuating, leading to a decline in the overall quality of life.

The impact on children is particularly profound. Witnessing or experiencing violence at a young age can have devastating and long-lasting effects on their emotional and psychological development. Children exposed to violence can develop post-traumatic stress disorder (PTSD), anxiety, depression and other mental health problems. Their academic performance may suffer and their social skills may be impaired. In addition, children who grow up in a violent environment are more likely to become violent themselves in the future, perpetuating the cycle of violence across generations. The cumulative effect on children's development results in significant social costs, including increased need for mental health services and increased risk of delinquency and criminal behaviour.

The societal impact of violence extends even further, affecting the very structure and stability of governments. In extreme cases, protracted cycles of violence can lead to the collapse of governing institutions, creating a vacuum that can be filled by criminal organisations or extremist groups. The erosion of trust in government further exacerbates the problem, leading to a decline in civic participation and undermining the legitimacy of political processes. The loss of social cohesion can lead to a decline in the quality of governance, making it more difficult to address the root causes of violence and implement effective solutions. This creates a dangerous cycle in which violence undermines governance, leading to further instability and violence.

Addressing the societal impact of violence requires a multi-pronged approach. Investing in community-based violence prevention programmes, strengthening law enforcement and the justice system, and providing comprehensive support services for victims are all critical steps. But it is equally important to address the underlying social and economic factors that contribute to violence, such as poverty, inequality and lack of opportunity. Creating opportunities for education, employment and social mobility can significantly reduce the likelihood of violence. Promoting social inclusion, fostering dialogue and understanding between different groups and strengthening community networks are also essential components of an effective strategy.

The role of cultural norms and values

The pervasive influence of culture on human behaviour is undeniable, and its impact on the prevalence of violence is particularly significant. While individual psychological factors and socio-economic conditions undoubtedly contribute to acts of violence, understanding the role of cultural norms and values is crucial to developing effective strategies for violence prevention and intervention.

A prominent aspect of this relationship is the cultural construction of masculinity. In many societies, particularly those with strong patriarchal structures, masculinity is often equated with aggression, dominance and the capacity for physical violence. Boys and men are often socialised to suppress emotions such as fear and sadness, while aggression and assertiveness are often seen as positive attributes, even celebrated as signs of strength and masculinity. This socialisation process can create an environment in which violent behaviour is not only tolerated, but in some cases encouraged and rewarded. The media, particularly popular culture, often reinforces this narrative by portraying violent characters as heroic or attractive, further normalising violent behaviour. This is not to say that all men from cultures that emphasise such definitions of masculinity are inherently violent; rather, it points to a social climate that can inadvertently foster an environment conducive to aggression.

The concept of "honour culture" provides another compelling example of how cultural values can directly influence the propensity to violence. In honour cultures, the reputation and social standing of an individual or family are of paramount importance. Any perceived slight or insult, real or imagined, is seen as a serious affront that demands immediate retribution. This often takes the form of violent acts, ranging from verbal threats to physical assault and even murder, to restore the victim's damaged honour. These actions are not necessarily driven by personal animosity, but rather by the overarching cultural imperative to maintain and defend one's honour. The social pressure to respond violently to perceived insults can be incredibly powerful, creating a cycle of violence that is deeply ingrained in the fabric of the culture. This cultural context makes violence not only acceptable but expected, even necessary, in certain situations.

Closely related to the culture of honour is the concept of revenge, which is a powerful motivator for violence in all cultures. While the desire for retribution is arguably a universal human emotion, its expression and acceptance varies greatly according to cultural norms. In some societies, revenge is seen as a legitimate and even morally justifiable response to harm or wrongdoing. This perspective often fosters a culture of retaliation, where acts of violence are met with counter-violence, creating a never-ending cycle of revenge. This cycle can be incredibly destructive, trapping communities in feuds and conflict for generations. The cultural acceptance of revenge can severely hamper efforts to resolve conflicts peacefully and promote reconciliation. The social normalisation of revenge often overshadows the importance of justice systems and peaceful conflict resolution strategies.

Furthermore, the influence of religious and ideological beliefs on violence cannot be overlooked. While many religions preach peace and non-violence, certain interpretations or radicalisations of these beliefs have historically been associated with violence and conflict.

Extremist groups often use religious or ideological rhetoric to justify violent actions, framing them as acts of faith or holy war. This manipulation of religious or ideological principles can provide a powerful motivation for violence, creating a sense of righteousness and moral justification that transcends conventional ethical considerations. Understanding how extremist groups use religious and ideological narratives to legitimise violence is essential to countering their influence and preventing the spread of violent extremism. The psychological impact of such beliefs, the sense of belonging to a group that is perceived to be fighting a just cause, can be particularly powerful.

The impact of cultural norms and values on violence is not limited to specific regions or communities. Even within seemingly homogeneous societies, variations in cultural practices can lead to significant differences in the prevalence of violence. Understanding the role of cultural norms and values in perpetuating violence requires a nuanced and culturally sensitive approach. Simple generalisations or sweeping judgments about particular cultures or communities are counterproductive. What is needed is a deeper understanding of the specific cultural factors that influence violence in different contexts. This requires engaging with communities and incorporating their perspectives and insights into violence prevention strategies.

Tackling this complex issue requires a multifaceted approach that includes both individual-level interventions and broader societal changes. Education plays a crucial role in challenging harmful cultural norms and promoting peaceful conflict resolution skills. This includes encouraging critical thinking about gender roles, addressing misconceptions about honour and revenge, and promoting religious tolerance and understanding. It also requires challenging the glorification of violence in the media and popular culture.

In addition, promoting economic opportunity and reducing social inequality can mitigate some of the underlying factors that contribute to violence in certain communities. Economic hardship and lack of opportunity can fuel frustration and resentment, making individuals more vulnerable to violence. Therefore, creating a sense of hope and ensuring equal opportunities can make a significant contribution to efforts to reduce violence.

Legislation and law enforcement also play a crucial role in addressing violence, but their effectiveness is greatly enhanced when implemented in conjunction with efforts to change cultural norms and values. Tougher laws and increased enforcement are often necessary, but not sufficient, to address the deep-seated cultural issues that contribute to violence. This must be seen in the wider context of societal change, including education, economic empowerment and community initiatives.

Effective violence prevention requires a comprehensive, long-term strategy that goes beyond treating symptoms. It requires a fundamental shift in cultural attitudes and beliefs, coupled with sustained efforts to promote peace, justice and equality in society. This process requires a collaborative effort between researchers, policy makers, community leaders and individuals committed to fostering a culture of peace. Ultimately, the creation of a violence-free society requires a deep and lasting transformation of cultural norms and values that either condone or normalise violence. This transformation requires addressing issues of social justice, gender equality and economic opportunity in order to truly mitigate the root causes of violence and create a more just and peaceful society. Only then can the complex interplay between cultural values and violence be effectively addressed, leading to a meaningful reduction in community violence.

Ideological conflict and community violence

Ideological conflict is fundamentally about competing visions of the world, different beliefs about how society should be organised, and conflicting values about justice, equality and the distribution of power. These differences, when deeply entrenched and perceived as an existential threat, can easily transcend rational discourse and erupt into violence. The intensity of such conflicts is heightened by the inherent human need for belonging and identity. Individuals often find comfort and validation in their ideological affiliations, and challenges to these beliefs can evoke strong emotional responses, including anger, fear and resentment. This emotional intensity can be easily manipulated and exploited by those who seek to incite violence.

The historical record is replete with examples of ideological conflict erupting into community-wide violence. Europe's centuries-long wars of religion are a stark reminder of the devastating consequences of religious intolerance. The Thirty Years' War, for example, was not simply a conflict over territory, but a brutal struggle between Catholics and Protestants, fuelled by deeply held religious beliefs and perceptions of the other as heretics. These religious differences were woven into the fabric of social life, shaping identities and allegiances and ultimately leading to widespread violence and social disruption. The sheer scale of the death and destruction caused by these wars is a sobering illustration of the destructive potential of ideological conflict.

Similarly, the rise of nationalism in the 19th and 20th centuries witnessed numerous instances of violent conflict driven by competing national identities and aspirations for self-determination. The Balkan Wars, for example, highlighted the destructive power of ethnic nationalism as competing national groups clashed over territory, resources and political dominance. The violence was characterised not only by inter-state warfare, but also by widespread atrocities and ethnic cleansing against civilian populations. These conflicts underlined the devastating consequences of unchecked nationalism and its potential to become a powerful catalyst for violence.

Contemporary examples of ideological conflict fuelling communal violence are readily apparent. The rise of extremist groups, whether driven by religious, political or ethnic ideologies, poses a significant threat to global peace and security. These groups often use violence to achieve their goals, targeting civilians and perpetuating cycles of violence and retaliation. The Islamic State of Iraq and Syria (ISIS), for example, has used a radical interpretation of Islamic theology to justify its campaign of violence and terror, recruiting members from diverse backgrounds and spreading its influence across multiple regions. The group's actions, including mass killings, beheadings and the destruction of cultural heritage sites, highlight the catastrophic consequences of extreme ideologies.

The role of social media and the Internet in amplifying and spreading extremist ideologies cannot be overlooked. The ease with which extremist groups can reach and recruit new members through online platforms exacerbates the problem. The anonymity afforded by the Internet, combined with sophisticated propaganda techniques, allows these groups to cultivate a sense of community and belonging among their followers, making them more susceptible to violence. The rapid spread of misinformation and hate speech online further polarises societies and fuels resentment against different ideological groups.

The impact of ideological conflicts is not limited to large-scale wars and terrorist attacks. Even within seemingly peaceful societies, simmering ideological tensions can erupt into community-level violence. Protests and demonstrations, while often the expression of legitimate grievances, can escalate into violence when fuelled by deep-seated ideological divisions. The presence of extremist elements within these protests can further exacerbate tensions, leading to confrontations with law enforcement and violent clashes between different groups. Dealing with such events requires a nuanced approach that addresses the underlying ideological conflicts while maintaining public order and preventing further escalation.

Addressing ideological conflicts and mitigating their potential for violence requires a multi-pronged strategy. Promoting dialogue and understanding between different groups is essential to foster tolerance and bridge ideological divides. Education plays a crucial role in promoting critical thinking, challenging extremist narratives and fostering respect for different perspectives. This includes promoting media literacy to counter the spread of misinformation and hate speech online. In addition, strengthening democratic institutions and promoting good governance can create more inclusive and equitable societies and reduce the grievances that can fuel ideological conflict.

However, dialogue and education alone are often not enough to address deep-seated ideological conflicts. In cases where violence has already erupted, robust conflict resolution mechanisms, including mediation, negotiation and arbitration, may be needed to facilitate the peaceful resolution of disputes. These mechanisms should involve all relevant parties and seek to address the underlying causes of the conflict. In some cases, external intervention, either by international organisations or neighbouring states, may be necessary to facilitate peace-building efforts.

Preventing violence stemming from ideological conflict also requires strong law enforcement and judicial systems. These systems should be able to deal effectively with acts of violence and hold perpetrators to account. But it is equally important to ensure that these systems are fair and impartial and do not discriminate against any particular group. A justice system that is perceived as biased can further exacerbate tensions and fuel resentment, ultimately undermining efforts to promote peace.

Finally, addressing the root causes of ideological conflict requires a long-term commitment to promoting social justice, economic opportunity and political inclusion. Inequality, marginalisation and lack of opportunity can provide fertile ground for extremist ideologies to flourish. Addressing these underlying issues is essential to fostering more inclusive and equitable societies that are less prone to violence fuelled by ideological divides. It will take a concerted effort by governments, civil society organisations and the international community to effectively address these complex issues. The path to peace is paved by promoting dialogue, tolerance, justice and equality. Only by addressing the root causes of ideological conflict can we hope to mitigate the devastating consequences of community-wide violence.

Media influence and the glorification of violence

The sheer volume of violent content readily available on various media platforms - from Hollywood blockbusters and graphic video games to news reports of real-world atrocities - raises serious concerns. While not all exposure to media violence necessarily leads to violent behaviour, research suggests a correlation between high levels of consumption and an increased likelihood of aggression, particularly among vulnerable individuals. This is not to argue for a simplistic cause and effect relationship, but rather to acknowledge the complex interplay between media influence and behavioural outcomes. The impact varies considerably depending on factors such as individual personality, pre-existing vulnerabilities and the specific context of media consumption.

A crucial aspect of media influence is its ability to desensitise individuals to the consequences of violence. Repeated exposure to graphic depictions of aggression, often without significant consequences for the perpetrators, can lead to a diminished emotional response to violence in real life. This desensitisation can manifest itself in reduced empathy for victims and reduced perception of the severity of violent acts. This effect is particularly strong when violence is presented as glamorous, exciting or even justified within the narrative. Action films, for example, often portray violence as a means to an end, a way of achieving justice or protecting loved ones, without fully exploring the devastating consequences for victims and communities.

This can subtly shift societal perceptions, making violence seem less reprehensible and more acceptable in certain circumstances.

Video games, in particular, have been the subject of intense debate about their potential to promote aggression. Many popular games feature interactive violence, requiring players to engage in acts of aggression in order to progress through the game. While the debate about causality is ongoing, some studies suggest a link between extensive exposure to violent video games and increased aggressive behaviour, particularly in individuals with pre-existing tendencies towards aggression. The interactive nature of these games, combined with the potential to reward and reinforce violent actions, may contribute to the normalisation of violence and the development of aggressive behaviour. Furthermore, the immersive nature of modern video games, with their high quality graphics and realistic sound effects, can heighten the emotional impact of violent scenarios, potentially exacerbating their influence on the player.

The pervasiveness of violence on television also deserves attention. While news reports play an important role in informing the public about acts of violence, repeated exposure to such images, often presented with little context or analysis, can contribute to a climate of fear and anxiety. Moreover, the portrayal of violence in fictional television programmes, often sanitised and devoid of realistic consequences, can reinforce the normalisation of violence and desensitise viewers to its harmful effects. The constant barrage of violent content, regardless of genre, may cumulatively contribute to a general societal tolerance of violence.

Furthermore, the increasing prevalence of reality television, which often features conflict and aggression as central plot points, raises further concerns about the normalisation of violence in entertainment. Beyond the direct impact on individual attitudes and behaviour, the media also play a crucial role in shaping societal perceptions of violence.

The way violence is portrayed - its justification, its consequences, the perpetrators and victims involved - influences public opinion and informs policy debates. For example, the media's portrayal of crime and punishment can shape public support for tougher sentencing, even when such policies have not been shown to be effective in reducing crime rates. Similarly, the media's focus on particular types of violence, such as gang violence or domestic violence, can influence the allocation of resources to prevention and intervention programmes. The media's selective focus, however unintentional, can distort the reality of violence and hinder effective policy development.

The influence of the media is not monolithic; its impact is mediated by a variety of factors, including individual differences, cultural context and social environment. It's therefore crucial to take a nuanced approach to understanding the relationship between media consumption and violent behaviour. Simple causal links are often misleading; instead, researchers emphasise the complex interplay of multiple factors that contribute to the development and expression of violence. However, the potential of the media to shape attitudes towards violence, to desensitise individuals to its consequences and to normalise aggression remains a major area of concern.

Addressing this issue requires a multi-pronged approach. Media literacy education plays a crucial role in empowering individuals to critically analyse media messages, identify bias and understand the potential impact of violent content. This includes equipping individuals with the skills to identify propaganda and manipulative techniques often used to glorify violence, and to cultivate a more discerning approach to media consumption.

The media industry also has a responsibility to promote responsible content creation. This includes carefully considering the potential impact of violent content, avoiding gratuitous depictions of violence, and striving for balanced and nuanced portrayals that accurately reflect the consequences of violent acts. Industry self-regulation, combined with government oversight and public pressure, can encourage responsible practices.

In addition, the creation of media that explicitly challenges the normalisation of violence and promotes alternative narratives is crucial. Positive role models, stories of resilience and conflict resolution, and media that focus on restorative justice can provide counter-narratives and counter the desensitisation fostered by violent content.

Finally, it remains essential to address the societal factors that contribute to the acceptance and normalisation of violence. This includes addressing issues such as poverty, inequality and social exclusion, which can create environments in which violence is more likely to occur. By addressing the root causes of violence in society, we can create a context in which the influence of the media is less likely to have harmful consequences. A holistic strategy combining media literacy, responsible media production and broader societal reform is needed to mitigate the harmful effects of media portrayals of violence. The challenge is to promote a media environment that informs and entertains without contributing to the normalisation and glorification of violence. This requires a collective effort by media producers, educators, policy makers and the public itself.

Hate crime and bias-motivated violence

Building on the previous discussion of the media's influence on the normalisation of violence, we now turn to a particularly insidious form of violence: hate crimes and bias-motivated attacks. These acts are not just isolated incidents of aggression; they represent a profound societal failure to uphold the principles of equality and respect for human dignity. They are fuelled by prejudice, discrimination and entrenched intolerance against individuals or groups on the basis of their race, ethnicity, religion, sexual orientation, gender identity, disability or other characteristics. Understanding the complex interplay of factors that contribute to these heinous acts is crucial to developing effective strategies for prevention and intervention.

The very definition of a hate crime highlights its specific nature. It goes beyond simple assault or vandalism to encompass acts of violence or intimidation motivated by prejudice against a victim's perceived membership of a particular group. This prejudice is the crucial differentiating factor that elevates the crime beyond a mere act of individual aggression to an attack on the very fabric of a tolerant and inclusive society. The motivation behind the crime - the perpetrator's prejudice - is as critical as the act itself, signalling an intention to harm not just an individual but an entire community.

The roots of hate crime lie in entrenched societal prejudices and discriminatory practices. These prejudices are often learned and reinforced through socialisation processes that include family upbringing, peer groups and exposure to societal stereotypes and prejudices. The media, as discussed above, play an important role in shaping perceptions and perpetuating harmful stereotypes, inadvertently fostering a climate in which hatred and intolerance can flourish. Education systems, by failing to adequately address issues of diversity and inclusion, can also inadvertently contribute to the normalisation of prejudice.

In addition, historical contexts and current socio-political events often exacerbate existing tensions and create fertile ground for hate crimes. Periods of social unrest, economic hardship and political polarisation often see a spike in hate-motivated violence, as fears and frustrations are channeled into attacks against marginalised groups. Scapegoating, the process of blaming a particular group for wider societal problems, becomes a common tactic, fuelling discriminatory attitudes and behaviour.

The impact of hate crime goes far beyond the immediate victim. These attacks cause profound psychological trauma and foster feelings of fear, vulnerability and insecurity within entire communities. A sense of safety and belonging is undermined, creating a climate of suspicion and mistrust. In addition, hate crimes send a chilling message of intolerance, deterring members of the targeted groups from participating fully in public life and potentially undermining social cohesion.

The legal response to hate crime varies from jurisdiction to jurisdiction, but generally hate crime legislation aims to increase penalties for bias-motivated crimes. These increased penalties reflect a societal understanding that hate crimes are not simply individual acts of violence, but attacks on the principles of equality and justice. However, tougher penalties alone are not enough to address the root causes of hate-motivated violence. A more comprehensive approach is needed, involving a multi-pronged strategy of law enforcement, education, community engagement and social reform.

Law enforcement has a crucial role to play in effectively investigating and prosecuting hate crimes. This requires specialised training for law enforcement officers to identify bias-motivated crimes, collect appropriate evidence and ensure sensitive and respectful treatment of victims. In addition, building trust between law enforcement and affected communities is essential to encourage victims to report hate crimes without fear of retaliation or inadequate response.

Education plays a crucial role in combating prejudice and promoting tolerance. Comprehensive education programmes, starting at an early age, are essential to foster critical thinking, empathy and understanding of diversity. These programmes should actively challenge stereotypes and prejudices and promote inclusiveness and respect for all individuals, regardless of their background. In addition, adult education initiatives that promote tolerance and challenge entrenched prejudices are equally important.

Community involvement is also essential. Collaboration between community organisations, law enforcement and government agencies can create safe spaces and support systems for victims of hate crime. Community-led initiatives that focus on reconciliation and dialogue can help bridge divides and promote understanding between different groups. These initiatives can include educational workshops, community dialogues and initiatives to promote positive intergroup relations.

Finally, it is essential to address the underlying social conditions that contribute to prejudice and discrimination. This requires tackling systemic inequalities and addressing issues such as poverty, unemployment and lack of access to education and health care. By creating a fairer and more just society, we can reduce the underlying tensions that often fuel hate crime. Tackling economic inequalities and promoting social mobility are integral parts of a long-term strategy.

Tackling hate crime and bias-motivated violence is an ongoing process that requires sustained commitment from individuals, communities and governments alike. It requires a holistic approach that addresses both the immediate consequences of hate crime and the deeper societal roots of prejudice and intolerance.

Cultural interventions to reduce violence

Cultural interventions offer a promising avenue for violence reduction by addressing the root causes of violence within specific community contexts. Unlike top-down approaches, which often ignore cultural nuances, these interventions use existing cultural values, beliefs and practices to promote peacebuilding and conflict resolution. The effectiveness of such interventions depends on a deep understanding of the target community's unique cultural landscape, including its social structures, power dynamics and communication styles. Conversely, a culturally insensitive approach risks alienating the community and undermining the impact of the intervention.

A crucial aspect of culturally sensitive interventions is identifying and working with community leaders and respected figures. These individuals often have considerable influence and can play an important role in disseminating violence prevention messages and promoting positive change. Their support and active participation is essential to build trust and ensure acceptance of the programme within the community. For example, in communities with strong religious affiliations, religious leaders can be powerful advocates for non-violent conflict resolution, using religious teachings and practices to promote peace and reconciliation. Similarly, community elders, who are often repositories of traditional wisdom and conflict resolution mechanisms, can be crucial partners in designing and implementing culturally appropriate interventions.

Successful interventions often involve adapting existing conflict resolution mechanisms to the specific community context. Many cultures have traditional ways of dealing with disputes and resolving conflicts peacefully. These methods, often based on community consensus and restorative justice, can provide a valuable foundation for violence reduction programmes. By building on existing cultural practices, interventions can foster a sense of ownership and participation, increasing their effectiveness.

Individual trauma and its long-term effects

The profound and lasting effects of violence go far beyond the immediate physical injuries. Victims often struggle with a complex set of long-term physical, psychological and emotional consequences that significantly affect their well-being, relationships and overall quality of life. Understanding these lasting effects is critical to developing effective support systems and interventions to promote healing and recovery.

One of the most significant long-term consequences of violence is the development of post-traumatic stress disorder (PTSD). PTSD is a debilitating mental health condition characterised by persistent intrusive memories, flashbacks, nightmares and avoidance related to the traumatic event. Victims of violence may experience intense anxiety, hypervigilance and difficulty sleeping, leading to significant impairment in daily functioning. The severity of PTSD symptoms can vary widely, depending on the nature and severity of the violence experienced, as well as individual factors such as pre-existing mental health conditions and social support networks.

In addition to PTSD, violence can lead to a range of other mental health problems. Depression, characterised by persistent sadness, loss of interest in activities and feelings of hopelessness, is a common consequence of experiencing violence. The trauma of violence can disrupt an individual's sense of safety and security, leading to feelings of helplessness, vulnerability and diminished self-worth. Anxiety disorders, including generalised anxiety disorder and panic disorder, can also develop as a result of violence, manifesting as excessive worry, fear and physical symptoms such as rapid heartbeat and shortness of breath. The chronic stress of living with the consequences of violence can overwhelm an individual's coping mechanisms, leading to a cascade of mental health problems that significantly affect their overall well-being. For example, a child who witnesses parental violence may develop anxiety disorders, struggle to feel safe in their own home and develop a fear of conflict.

The physical effects of violence can also go beyond immediate injuries. Victims may experience chronic pain, headaches and gastrointestinal problems as a result of the physical trauma inflicted on them. These physical ailments can have a significant impact on daily functioning, limiting a person's ability to work, participate in social activities and maintain relationships. The ongoing physical pain and discomfort can lead to further psychological distress, exacerbating existing mental health problems and affecting overall quality of life. For example, a victim of violence who suffers a serious head injury may experience chronic headaches and cognitive difficulties that affect their ability to work and socialise.

Violence can also have a profound and long-lasting impact on a person's social relationships. The experience of violence can undermine trust and create feelings of isolation and alienation. Victims may find it difficult to form and maintain close relationships for fear of betrayal or further harm. The emotional burden of carrying the trauma of violence can strain existing relationships, leading to conflict and ultimately relationship breakdown. For example, a survivor of domestic violence may find it difficult to trust new partners for fear of repeating past patterns of abuse. This lack of trust can lead to isolation and difficulties in forming healthy relationships. Children who witness domestic violence may have difficulty forming healthy peer relationships, struggling to understand healthy boundaries and expressing their emotions effectively.

The impact of violence also extends to a person's general wellbeing. Victims may experience a decline in their overall physical and mental health, leading to a reduced quality of life. Difficulties sleeping, eating and concentrating can have a significant impact on a person's daily functioning and ability to engage in meaningful activities. Violence can lead to feelings of hopelessness and despair, undermining a person's motivation and ability to care for themselves.

The long-term effects of violence can be devastating, leading to significant impairment in many aspects of a person's life. For example, a person with chronic PTSD following a violent assault may find it difficult to maintain employment, leading to financial hardship and exacerbating their emotional distress. The overall impact on their wellbeing becomes a cycle of negative consequences, highlighting the urgency of timely and comprehensive support.

The recovery process is unique to each individual and requires personalised approaches tailored to the specific needs and circumstances of the victim. The severity of the trauma, pre-existing mental health conditions, the availability of support systems and the resilience of the individual all contribute to the trajectory of recovery. Some individuals may recover relatively quickly, while others may struggle with the long-term effects of violence for many years. However, with appropriate support and intervention, most victims can successfully navigate their way to healing and rebuilding their lives. Ongoing support, even after initial symptoms have subsided, is crucial to preventing relapse and maintaining long-term stability. It is important to remember that recovery is not a linear process and setbacks are common. Continued access to resources, both professional and social, is essential to ensure sustainable healing and recovery.

In conclusion, the impact of violence goes far beyond the immediate physical and emotional trauma. Victims often experience a complex range of long-term physical, psychological and social challenges that can significantly affect their quality of life.

However, effective interventions, including early intervention, access to mental health services, social support and trauma-informed care, can significantly improve recovery rates and mitigate the lasting effects of violence. A holistic approach that addresses both the immediate and long-term needs of victims is essential to promote healing, build resilience and empower survivors to rebuild their lives. The continued development and expansion of comprehensive support systems remains essential to ensure that all victims of violence have access to the assistance they need to overcome the challenges they face and achieve sustainable recovery.

The economic cost of violence

The devastating human toll of violence goes far beyond individual suffering; it imposes a significant and often overlooked economic burden. The costs associated with violence are multifaceted, encompassing a wide range of expenses borne by individuals, families and society as a whole. Understanding this economic impact is critical to developing effective prevention strategies and efficiently allocating resources to address this pervasive societal problem.

One of the most direct and easily quantifiable costs is health care. Victims of violent crime often require extensive medical care, including emergency room visits, hospitalisation, surgery, rehabilitation and ongoing treatment for injuries ranging from minor bruises and lacerations to life-threatening trauma. The severity of the injuries determines the level and cost of care, with severe cases requiring prolonged hospitalisation, specialised surgery and extensive rehabilitation, resulting in exorbitant medical bills that place a significant burden on both individuals and the healthcare system.

In addition to direct health care costs, violence results in significant indirect economic losses due to lost productivity. Victims of violence may be unable to work due to physical injuries, mental health problems or the need for extensive recovery time. This loss of income affects not only the victim but also their families, leading to financial instability and hardship. The impact goes beyond the victim's immediate loss of earnings. Businesses may experience lost productivity due to employee absenteeism, the need to replace injured workers, and the potential disruption of operations following acts of violence on their premises. The knock-on effects extend to the wider economy, with reduced economic output and national productivity as a result of violence-related absenteeism. Long-term disabilities resulting from violent attacks can exacerbate these economic losses, requiring ongoing medical care and potentially leading to permanent loss of earning capacity.

The justice system also bears a significant economic burden in responding to violence. Law enforcement agencies incur substantial costs associated with investigation, apprehension, prosecution and incarceration. These costs include personnel salaries, equipment maintenance, forensic analysis and court proceedings. The costs associated with lengthy trials, particularly in complex cases involving multiple victims or perpetrators, can be astronomical.

In addition, the costs associated with incarcerating convicted offenders, including the provision of food, shelter, health care and rehabilitation programmes, add to the overall economic burden. The length of the legal process and the length of prison sentences contribute significantly to the escalating financial burden on taxpayers. These resources could be better spent on prevention and community-based solutions to address the root causes of violence.

Beyond law enforcement and the justice system, the societal response to violence requires significant investment in social services. Victims of violence often require comprehensive support, including counselling, therapy, social work services and assistance with housing, job training and other basic needs. These services are crucial to the recovery and reintegration of victims into society, but they represent a significant financial commitment by both public and private organisations. The need for long-term support, particularly for victims of severe trauma, underlines the continuing economic burden of violence.

The economic costs of violence are not limited to the immediate aftermath. They extend to the long-term consequences and the need for ongoing support and intervention. Victims of violence may experience long-term disabilities that require ongoing medical care and support, placing a continuing economic burden on both individuals and the health care system. Similarly, the psychological consequences of violence, such as PTSD and depression, can lead to long-term health problems, requiring extensive mental health treatment and potentially affecting the victim's ability to maintain employment. These chronic conditions place a significant burden on health care resources and contribute to the overall economic impact of violence. In addition, fear of crime and violence can lead to lower property values, reduced tourism and stunted economic development in affected communities, creating a cycle of economic decline and social instability.

The development of a comprehensive economic model to quantify the total cost of violence requires the integration of these various components. Such a model would need to take into account both direct costs, such as health care and law enforcement expenditures, and indirect costs, such as lost productivity, reduced property values, and the costs associated with the broader social and economic consequences. The model would require a multi-faceted approach, incorporating data from a variety of sources, including health care providers, law enforcement, the judiciary, social service organisations and economic indicators. Collecting and integrating this data is a significant challenge, but is essential to accurately assess the true economic burden of violence and to inform policy decisions aimed at mitigating its impact. In addition, the model should incorporate long-term costs and projections to better understand the ongoing and accumulating financial burden of violence. The criminal justice system and its response to violence

The criminal justice system, a cornerstone of social order, plays a central role in responding to violence. However, its effectiveness is a complex and often debated issue. The system's approach to violence, encompassing law enforcement, prosecution and sentencing, has a significant impact on both individual lives and community safety. A critical examination reveals both strengths and inherent limitations in its ability to prevent, deter and punish acts of violence, ultimately shaping the societal consequences discussed above.

One of the primary functions of law enforcement in addressing violence is proactive crime prevention. This includes strategies such as community policing, where officers build relationships with residents, fostering trust and cooperation. Such initiatives aim to identify and address potential sources of conflict before they escalate into violence. However, the success of community policing depends on factors such as adequate officer training, resource allocation and the willingness of community members to engage with the police. In areas where police-community relations are strained by historical injustices or lingering prejudices, proactive strategies may fail and exacerbate existing tensions rather than fostering cooperation. Furthermore, even with robust community policing programmes, resource constraints may limit the system's ability to effectively patrol high-crime areas, leaving residents vulnerable.

The investigative aspect of policing is crucial to bringing perpetrators of violence to justice. Effective investigations require skilled detectives, forensic expertise and a commitment to thorough data collection and analysis. Technological advances, such as DNA analysis and digital forensics, have significantly improved law enforcement's ability to solve violent crimes. However, challenges remain. Backlogs in forensic analysis, understaffing and inadequate training can hamper the investigative process, leading to delays in bringing cases to trial and potentially affecting conviction rates. In addition, the collection and preservation of evidence requires strict adherence to legal standards and procedural protocols to ensure its admissibility in court and to minimise the possibility of wrongful convictions or acquittals.

The role of the prosecution is to ensure that those accused of violent crimes are held accountable under the law. This involves carefully building cases, presenting evidence in court and arguing for convictions. The success of prosecutions depends on factors such as the strength of the evidence, the skill of prosecutors and the cooperation of witnesses. However, prosecutorial discretion raises complex ethical and practical considerations. Overcharging, plea bargaining and decisions to pursue or dismiss charges can have a significant impact on the outcome of cases, potentially leading to inconsistent sentencing and reducing the overall effectiveness of the system. In addition, conscious and unconscious biases within the prosecution may inadvertently influence case selection and outcome, leading to disparities in sentencing based on factors unrelated to the seriousness of the offence.

Sentencing is the final stage in the criminal justice process where judges determine the appropriate punishment for convicted offenders. Sentencing philosophies vary, ranging from retribution (punishment for punishment's sake) to rehabilitation (aiming to reform offenders), incapacitation (removing offenders from society) and deterrence (discouraging future crime). The choice of sentence can have a profound impact on the individual and society. Long prison sentences, while apparently effective in incapacitating offenders, do not always deter future offending and may contribute to reoffending due to the lack of rehabilitation efforts. Alternative sentencing options, such as community service, probation and drug treatment programmes, can provide more effective avenues for rehabilitation and reintegration into society, thereby minimising the likelihood of future offending. However, these programmes often require significant resources and community support, which may not always be readily available, particularly in under-resourced communities.

The effectiveness of the criminal justice system's response to violence is further complicated by factors beyond its direct control. Underlying social conditions, including poverty, inequality, lack of educational opportunities and access to mental health services, contribute significantly to violent crime. The capacity of the system to address these root causes is limited. While rehabilitation and crime prevention programmes can help address individual factors, broader societal reforms are needed to effectively address the systemic issues that fuel violence. Without addressing these broader societal factors, the criminal justice system, however effective it may be in individual cases, is fighting an uphill battle.

Moreover, disparities in the application of justice based on race, ethnicity, socio-economic status and other factors undermine the legitimacy and effectiveness of the system. Studies consistently show that people from marginalised communities are disproportionately affected by the criminal justice system, facing harsher sentences and less access to resources than their more privileged counterparts. Addressing these systemic inequalities is critical to building a fairer and more effective justice system that truly addresses the root causes of violence and ensures equitable outcomes for all. This requires a thorough examination of policing practices, prosecutorial decisions and judicial sentencing to reduce implicit bias and systemic racism, thereby ensuring fairer treatment across all communities.

Another critical area is support for victims of violence. The justice system plays an important role in providing access to services such as victim support programmes, counselling and financial assistance. Effective support systems are essential to help victims recover from trauma, navigate the legal process and rebuild their lives. However, the availability and quality of victim support services can vary widely across regions and jurisdictions, with resource limitations affecting the effectiveness of the support provided. Moreover, the experience of navigating the justice system can itself be traumatic for victims, posing additional challenges to their healing and reintegration into their communities. Improving the victim support system is essential not only to ensure justice for victims, but also to promote confidence in the system as a whole.

Defining verbal violence and its scope

Verbal violence, as opposed to simple disagreement or assertive communication, involves the deliberate use of language to cause emotional, psychological or even physical harm. It's a complex phenomenon that goes far beyond simple name-calling or harsh words. The insidious nature of verbal violence lies in its ability to subtly undermine self-esteem, manipulate and control, often leaving lasting scars that are difficult to detect and even harder to heal. It goes beyond casual arguments and encompasses a wide range of behaviours, from overt insults and threats to more covert forms of manipulation and gaslighting. Understanding this nuanced spectrum is crucial to effectively recognising, addressing and ultimately preventing its harmful effects.

A key element that distinguishes verbal violence from assertive communication is the intent behind the words. Assertive communication, while potentially firm and direct, aims to express one's needs and opinions without demeaning or harming others. It focuses on clear and respectful communication, even when there is disagreement. Verbal violence, on the other hand, is driven by a desire to control, dominate or inflict pain. Words are specifically chosen to hurt, belittle or intimidate, often with the underlying aim of silencing or subjugating the victim. This intentionality distinguishes a sharp rebuke from a targeted verbal attack.

The forms of verbal violence are diverse and can manifest themselves in different ways. Direct attacks are the most obvious. These include explicit insults, derogatory remarks, threats of harm (physical or emotional) and aggressive, belittling language. Such attacks are often easily recognised as hurtful and malicious.

Examples include shouting obscenities, making direct threats ("I'll make sure you regret this") or using demeaning labels ("You're worthless", "You're incompetent"). These are often blatant and leave little room for misinterpretation. However, it's not just the intensity of the language that matters, but also the context and the relationship between the people involved.

A seemingly minor insult from a close family member can carry much more weight and inflict deeper wounds than a harsh criticism from a stranger.

However, the insidious nature of verbal violence is often revealed through its indirect forms. Passive aggression, a common tactic, masks aggression behind seemingly innocuous behaviour. Sarcasm, subtle put-downs, backhanded compliments and withholding affection or support are all examples of indirect verbal aggression.

These subtle acts can be difficult to recognise as abusive, often leaving victims questioning their own perceptions and feeling confused and disoriented. For example, the seemingly innocuous statement, "Oh, you tried your best, I guess," laced with sarcasm, can be more damaging than a direct insult, especially if it is repeated. Similarly, constant criticism disguised as helpful advice can erode a person's self-esteem over time.

Another key component is manipulative language. This is the use of words to control or influence the emotions and behaviour of others.

Gaslighting, a form of manipulation, involves making the victim question their own sanity and perception of reality. Examples include denying events that have happened, twisting words to blame the victim, and subtly undermining their confidence. A common example is a partner who consistently denies their abusive behaviour, causing the victim to doubt their own memories and feelings. This creates a cycle of confusion and dependency, making it harder for the victim to escape the abusive situation.

The scope of verbal abuse extends far beyond personal relationships. Workplaces are often hotbeds of subtle and overt verbal abuse. Workplace bullying, including harassment and intimidation, can have a significant impact on employees' mental health and job satisfaction. The power dynamics inherent in the workplace often amplify the effects of verbal abuse, with supervisors or managers having more power to inflict harm without fear of reprisal. Consider a manager who consistently belittles an employee's ideas in meetings, undermines their contributions or publicly humiliates them, damaging their professional reputation and self-worth. This kind of subtle but persistent abuse can have a devastating impact on the victim's career and overall wellbeing.

Educational institutions are not immune. Bullying, both verbal and physical, is a pervasive problem in schools at all levels. The repetitive nature of bullying, combined with the power imbalance between bully and victim, makes it particularly damaging. Children and adolescents in particular are vulnerable to the long-term psychological effects of verbal bullying, which can extend into adulthood. Imagine a child who is constantly subjected to abuse and insults from peers. This can lead to social isolation, reduced academic performance, anxiety and depression, potentially affecting their self-esteem and mental health for years to come.

The digital age has introduced a new dimension to verbal violence: cyberbullying. The anonymity and wide reach of online platforms can amplify the effects of verbal abuse. Cyberbullying often involves spreading rumours, threats, insults and hateful messages online, with the potential to reach a large audience and cause significant damage to the victim's reputation and emotional well-being. The relentless nature of online harassment, the ease with which hurtful messages can be spread, and the lasting digital footprint of abusive content contribute to its unique severity.

Statistics paint a disturbing picture of the prevalence and impact of verbal abuse around the world. While accurate figures are difficult to obtain due to under-reporting and varying definitions, studies consistently show a significant problem across different populations and settings. Research shows that verbal abuse has a significant impact on mental health, leading to increased rates of anxiety, depression, post-traumatic stress disorder (PTSD) and even suicidal thoughts. Long-term effects can include difficulty forming and maintaining healthy relationships, low self-esteem and reduced trust. In addition, verbal abuse is often linked to other forms of abuse, including physical and economic abuse, creating a complex cycle of harm.

Forms of Verbal Abuse - A Taxonomy

Forms of verbal abuse manifest themselves in a bewildering variety of ways, making identification and intervention difficult. To understand the scope of the problem, it's helpful to categorise these forms, recognising that these categories are not mutually exclusive and often overlap in practice. The impact of verbal abuse depends not only on the severity of individual incidents, but also on the cumulative effect of repeated patterns of behaviour, the context in which they occur and the power dynamics at play.

We can broadly divide verbal abuse into two main categories: direct attacks and indirect forms of aggression. Direct attacks are the most obvious forms of verbal abuse. They are overt expressions of hostility designed to deliberately cause pain or humiliation. They are characterised by their explicit and unambiguous nature, leaving little room for misinterpretation. This category covers a range of behaviours from insults and verbal abuse to explicit threats of violence or harm.

Insults are perhaps the most common form of direct verbal aggression. They are deliberately chosen words intended to demean, belittle or ridicule the target. This can range from simple verbal abuse ("stupid", "idiot") to more sophisticated and cutting remarks aimed at personal vulnerabilities or insecurities. The power of an insult lies in its ability to strike at the heart of a person's self-esteem, making them feel ashamed, embarrassed or inadequate. For example, a parent who consistently calls a child "useless" or "a disappointment" can have a profound and lasting negative effect on the child's self-esteem.

Threats, whether explicit or implied, are another important category of direct verbal aggression. These are statements that convey an intention to cause harm, whether physical, emotional or psychological. Threats can range from subtle hints of future harm ("You'll regret this") to explicit statements of violence ("I'm going to hurt you"). The seriousness of a threat is determined not only by its explicitness, but also by the context and the perceived credibility of the threatener. A threat made by someone with a history of violence carries significantly more weight and causes more fear than one made by someone known to be non-violent. Aggressive language, characterised by a hostile and controlling tone, also falls under the umbrella of direct assault. This includes shouting, interrupting, using a condescending tone and using language to dominate or intimidate the conversation. Aggressive language often creates a climate of fear and intimidation that prevents open and honest communication. For example, constantly interrupting someone during a conversation, talking over them or ignoring their points is a clear manifestation of aggressive communication designed to control the interaction.

Indirect forms of verbal abuse are more subtle and insidious, making them harder to recognise and address. These are instances of aggression hidden behind seemingly innocuous behaviour or words.

Passive aggression, a hallmark of indirect verbal abuse, involves expressing hostility indirectly, often through actions rather than direct confrontation. This can take the form of sulking, giving the silent treatment or subtly sabotaging the target's efforts. A common example is consistently failing to meet commitments or deadlines, causing stress and difficulty for others without explicitly expressing resentment.

Sarcasm, when used maliciously, can be a powerful weapon of indirect verbal abuse. While sarcasm can be used playfully, its deliberate use to hurt or humiliate is a form of verbal aggression. A backhanded compliment that appears positive on the surface but contains a hidden insult is another form. For example, saying "That's a nice dress for someone your age" is a backhanded compliment that undermines the target's appearance and self-worth.

Another common form of indirect verbal abuse is manipulation. This involves using language to control or influence another person's emotions and behaviour. This can range from subtle forms of guilt-tripping ("If you loved me, you would...") to more overt forms of emotional blackmail ("If you don't do this, I will..."). Gaslighting, a particularly insidious form of manipulation, involves making the victim question their own sanity and perception of reality. This often involves denying events that have happened, twisting words to blame the victim, and subtly undermining their confidence. A repeated pattern of such behaviour can lead to a profound sense of disorientation and self-doubt.

Withholding affection or support, while seemingly passive, can be a powerful form of indirect verbal abuse. The consistent lack of emotional support, empathy and positive reinforcement can significantly damage a person's self-esteem and mental health. For example, a partner who consistently refuses to offer encouragement or validation, even in the face of significant achievements, is engaging in a form of indirect verbal abuse.

The context in which verbal abuse occurs greatly influences its impact. For example, verbal abuse in intimate relationships can be particularly devastating because of the inherent power imbalance and expectation of support and trust. Children are particularly vulnerable to verbal abuse from parents or caregivers because they are completely dependent on these adults for their well-being. Power dynamics in the workplace can exacerbate the impact of verbal abuse, with supervisors or managers able to inflict harm with relatively little accountability. Similarly, the anonymity afforded by online platforms can embolden perpetrators of cyberbullying, leading to particularly relentless and damaging forms of verbal aggression.

Cultural norms and individual differences have a significant impact on the interpretation and experience of verbal abuse. Certain forms of language that may be considered acceptable or even humorous in one cultural context may be deeply offensive in another. Similarly, individual sensitivities and experiences shape how people perceive and respond to verbal interactions. What one person may dismiss as a minor offence may be perceived as a significant act of aggression by another. Therefore, understanding cultural and individual contexts is essential to accurately assess verbal behaviour.

The legal implications of certain forms of verbal aggression, such as hate speech, should also be considered. Hate speech, defined as language that attacks or degrades a person or group on the basis of race, religion, gender, sexual orientation or other protected characteristics, is often subject to legal sanctions in many jurisdictions. However, the line between offensive language and hate speech can be blurred, making legal intervention complex and challenging.

This exploration of forms of verbal abuse highlights the multifaceted nature of the problem. Understanding these different manifestations, their varying degrees of subtlety and the contexts in which they occur is crucial to effective prevention and intervention. The following chapters explore the psychological impact of verbal abuse, strategies for recognising and addressing it, and methods for promoting healthier communication patterns. The aim is to empower individuals to protect themselves and others from the devastating effects of verbal violence and to build a more compassionate and respectful society.

The psychology of verbal aggression

The psychology of verbal aggression is a complex interplay of individual factors, learned behaviours and social influences.

Understanding these mechanisms is crucial to the effective management and prevention of verbal violence. While the outward manifestation of verbal abuse may seem straightforward - a string of insults or a barrage of threats - the underlying motivations are often far more nuanced and deeply rooted in the perpetrator's psyche.

One prominent factor is anger management. Individuals with poor anger management skills often resort to verbal aggression as a maladaptive coping mechanism. Their inability to effectively regulate their emotions leads to outbursts of anger that manifest as verbal aggression. It's not just a case of 'losing their temper'; it's a pattern of behaviour that often stems from unresolved childhood trauma, learned responses to conflict, or underlying mental health conditions such as anxiety or depression. For example, someone who was often verbally abused as a child may unconsciously repeat this behaviour in their own relationships, perpetuating a cycle of violence. Therapy that focuses on emotional regulation and anger management techniques, such as cognitive behavioural therapy (CBT), can be very effective in addressing this root cause of verbal aggression.

Power dynamics play an important role in verbal aggression. Individuals who seek to exert control or dominance over others may use verbal abuse as a means of exerting power. This is particularly evident in relationships where there's a significant power imbalance, such as between an employer and employee, a parent and child, or a perpetrator and victim. The perpetrator uses words as a weapon to intimidate, humiliate and ultimately control the victim.

This dynamic is often seen in bullying behaviour, where the aggressor targets individuals who are perceived as weaker or less powerful. The repeated use of verbal abuse reinforces the power imbalance, silencing the victim and reinforcing the aggressor's dominance.

Understanding the power dynamics at play is essential in order to intervene and address the root causes of aggression, focusing on empowering the victim and challenging the perpetrator's control.

Low self-esteem is another important psychological factor associated with verbal aggression. Paradoxically, people with low self-esteem may resort to verbal abuse in order to increase their own self-worth. By putting others down, they try to raise themselves up, creating a false sense of superiority. This behaviour is often unconscious, driven by deep-seated insecurities and a need for validation. Aggression becomes a defence mechanism, protecting the individual from confronting their own inadequacies. Such individuals often have fragile egos, are hypersensitive to criticism and tend to overreact to perceived slights. Their verbal attacks become a desperate attempt to control their environment and protect themselves from perceived threats, even if these threats are entirely imagined. Therapy that focuses on building self-esteem and promoting healthier coping mechanisms is often a crucial component in treating this type of aggression.

Learned behaviour plays an important role in perpetuating cycles of verbal violence. Individuals who witnessed verbal abuse as children are more likely to engage in such behaviour themselves. This demonstrates how aggression can be transmitted across generations and become a normalised pattern of communication within a family or social group. Children learn through observation and imitation and internalise these patterns as acceptable ways of interacting and resolving conflict. Breaking this cycle requires conscious effort, intervention and the adoption of healthier communication strategies. Educational programmes that focus on healthy relationship dynamics and alternative conflict resolution methods can help prevent the transmission of these learned behaviours.

Social influences also contribute significantly to the prevalence of verbal aggression. Cultural norms and societal expectations can shape individuals' perceptions of acceptable behaviour. In some cultures, aggressive communication styles may be more accepted or even seen as a sign of strength or assertiveness. This normalisation of aggressive behaviour contributes to its persistence and acceptance within certain communities. In addition, exposure to aggressive media, such as violent video games or films, can desensitise individuals to the harmful effects of verbal abuse and contribute to its normalisation. Public awareness campaigns, media literacy programmes and the promotion of positive role models can help counteract these social influences and promote a culture of respect and healthy communication.

The relationship between verbal abuse and other forms of violence deserves careful consideration. Verbal aggression is often a precursor to physical violence, serving as an escalating step in a pattern of abusive behaviour. The normalisation of verbal aggression can desensitise both the perpetrator and the victim, leading to an increased tolerance of increasingly aggressive behaviour. This gradual escalation can be devastating, with verbal abuse acting as a gateway to more serious forms of violence.

Therefore, addressing verbal aggression at an early stage is crucial in preventing escalation to physical violence. Interventions need to focus not only on managing verbal aggression, but also on addressing the underlying issues that contribute to the potential for violence.

By recognising the complex interplay of these factors, we can better equip individuals, families and communities to build healthier, more respectful and safer patterns of communication. This knowledge empowers individuals to recognise, address and mitigate the harmful effects of verbal abuse, thereby fostering a more compassionate and understanding society.

The effects of verbal abuse on individuals and relationships

The insidious nature of verbal abuse goes far beyond the immediate sting of hurtful words. Its effects reverberate through the lives of individuals and relationships, leaving a trail of emotional, psychological and even physical damage. Understanding these consequences is crucial to recognising the seriousness of the problem and developing effective strategies for prevention and intervention.

One of the most immediate and profound effects of verbal abuse is the erosion of self-esteem. Repeated exposure to insults, belittling remarks and constant criticism undermines a person's sense of self-worth. Victims begin to internalise the negativity, questioning their abilities, intelligence and even their value as human beings. This can lead to a cycle of self-doubt and depression, making it difficult for them to form healthy relationships and pursue their goals. The constant barrage of negative messages can fundamentally alter an individual's sense of self, leaving them feeling helpless, inadequate and unworthy of love and respect. This damage is particularly acute for children who are still developing their sense of self, as the constant verbal abuse can severely affect their self-confidence and future relationships. The effect on trust is equally devastating. Verbal abuse destroys the foundation of trust in relationships, both romantic and platonic.

When someone consistently uses words as weapons, it creates an atmosphere of fear and insecurity. The victim learns to anticipate attacks and is constantly bracing for the next onslaught of negativity. This constant state of hyper-vigilance undermines their ability to relax and feel safe in the relationship. Trust, a cornerstone of any healthy relationship, becomes impossible to rebuild when words are used to manipulate, control and demean. The betrayal of trust caused by verbal abuse can lead to deep-seated anxiety and make it extremely difficult for the victim to form trusting relationships in the future. Fear of repetition often prevents victims from voicing their concerns or seeking help, further isolating them and exacerbating the damage.

The emotional wellbeing of victims is profoundly affected by verbal abuse. The constant negativity and stress can manifest itself in a variety of ways, including anxiety, depression, post-traumatic stress disorder (PTSD) and even suicidal thoughts. The emotional toll is significant and affects every aspect of the victim's life. Simple daily tasks can become overwhelming, and maintaining healthy relationships with others becomes increasingly difficult. Feeling constantly under attack can lead to a profound sense of loneliness and isolation, even within seemingly close relationships. The cumulative effect of emotional distress can have a significant impact on physical health, leading to increased susceptibility to illness and reduced overall well-being. This constant emotional turmoil can affect sleep patterns, appetite and general physical health, creating a vicious cycle of emotional and physical distress.

Beyond the individual, verbal abuse has a devastating effect on interpersonal relationships. In romantic relationships, it creates an environment of fear, control and power imbalance. The abuser uses words to dominate and manipulate their partner, undermining their self-esteem and isolating them from support networks. This can escalate to physical violence, creating a highly dangerous situation for the victim. Even in seemingly less severe cases, the constant negativity and emotional turmoil can severely damage the relationship, leading to separation, divorce or prolonged periods of unhappiness. The lack of healthy communication and the presence of constant conflict prevent the development of mutual respect and understanding, essential components of any stable partnership.

Similarly, verbal abuse has a significant impact on platonic relationships. Friendships are based on trust, respect and mutual support. When verbal abuse enters the picture, this foundation crumbles. Victims may become emotionally distant from their friends, fearing judgement or further negativity. The constant stress and emotional exhaustion can make it difficult to maintain healthy social relationships, further isolating the individual and exacerbating the negative effects of the abuse. Fear of being judged or misunderstood often prevents victims from disclosing the abuse, which hinders their ability to receive support and understanding from their social circle.

Children who witness or experience verbal abuse suffer profound and lasting effects. Children are particularly vulnerable to the effects of verbal abuse because their sense of self and understanding of relationships are still developing. Witnessing verbal abuse between parents or caregivers can lead to emotional distress, anxiety and behavioural problems. Children who are the direct targets of verbal abuse can experience significant developmental delays, low self-esteem and difficulty forming healthy relationships later in life. They may develop unhealthy coping mechanisms, such as aggression or withdrawal, which affect their social and academic performance. Long-term effects can include struggles with depression, anxiety and difficulties forming healthy relationships in adulthood. Early intervention and support is essential to mitigate these devastating effects.

The long-term effects of verbal abuse can be far-reaching and deep-seated. Victims can struggle with anxiety, depression and PTSD for years after the abuse has stopped. Damage to self-esteem and trust can remain, affecting their ability to form healthy relationships and achieve their full potential. The emotional scars can be deep and may require extensive therapy and support to heal. The experience of verbal abuse can lead to difficulties in forming trusting relationships, managing emotions effectively and setting healthy boundaries. The impact can extend to work performance, financial stability and overall life satisfaction.

It is important to stress the importance of seeking help and using the resources available. If you or someone you know is experiencing verbal abuse, there are people who can help.

Many organisations provide support, advice and resources for victims of abuse. These resources can provide a safe space to process emotions, develop coping mechanisms, and create a safety plan. Seeking professional help is not a sign of weakness; it's a brave step towards healing and reclaiming your life. Help is available and recovery is possible.

Verbal abuse in different contexts

The devastating effects of verbal violence are not limited to the intimate sphere of personal relationships. Its insidious reach extends into the diverse tapestry of human interaction, manifesting itself differently in different contexts, each with its own unique power dynamics and vulnerabilities. Understanding these variations is critical to developing effective strategies for prevention and intervention in each specific setting.

Families, the supposed bedrock of support and love, can unfortunately become a breeding ground for verbal violence. The inherent power imbalances within family structures, often rooted in age, authority or perceived dependency, can exacerbate the harm caused by abusive language. For example, parents may use belittling remarks or harsh criticism to control or manipulate their children, subtly undermining their self-esteem and creating a cycle of fear and insecurity. Similarly, siblings, driven by jealousy, rivalry or a desire to dominate, may engage in verbal aggression that causes significant emotional distress and affects their relationship. The long-term consequences of this verbal abuse within the family can manifest in difficulties forming healthy relationships, low self-esteem and increased vulnerability to other forms of abuse later in life. The insidious nature of this abuse lies in its normalisation within the family unit - hurtful words become accepted as part of everyday communication, making it harder for victims to recognise the abuse and seek help. The perceived sanctity of the family unit often shields the abuse from outside scrutiny, leaving victims isolated and vulnerable.

The workplace, often thought of as a professional and respectful environment, can also become a battleground for verbal abuse. Workplace bullying, often characterised by repeated verbal attacks, humiliation and intimidation, can have a serious impact on an employee's mental and physical health, productivity and job satisfaction. The power dynamics of the workplace, particularly its hierarchical structure, can make employees particularly vulnerable to verbal abuse from supervisors or colleagues. Those in lower positions may fear retaliation or job loss if they report abusive behaviour.

The subtle nature of verbal abuse in the workplace, such as constant criticism, exclusion from meetings or spreading rumours, can be difficult to document and prove, making effective intervention difficult. The constant stress and anxiety associated with verbal abuse in the workplace can lead to burnout, absenteeism and even sick leave. This in turn can affect overall productivity and morale in the workplace, creating a toxic environment for all. The economic consequences of such abuse can be profound, with individuals losing their jobs, affecting their financial stability and hindering their career progression. In addition, the stigma associated with reporting such abuse often prevents individuals from seeking help or redress.

Schools, supposedly places of learning and growth, are not immune to the pervasive nature of verbal violence. Bullying in its various forms, including verbal abuse, is a significant problem in schools and leaves its victims emotionally scarred. Children and adolescents are particularly vulnerable to verbal attacks because their developmental stage makes them more susceptible to the psychological impact of hurtful words. Peer dynamics in schools, particularly the power imbalances created by social hierarchies and cliques, can create an environment in which verbal abuse thrives.

Cyber bullying, a relatively new form of verbal violence, has significantly exacerbated the problem by extending the reach of verbal abuse beyond the confines of the schoolyard. The anonymity afforded by the internet often emboldens bullies, leading to more intense and pervasive forms of verbal abuse. The constant harassment and ridicule experienced by victims can have a devastating effect on their self-esteem, academic performance and general mental health. Long-term effects can include anxiety, depression and even suicidal thoughts, highlighting the urgent need for effective anti-bullying programmes and initiatives in schools.

The often-overlooked bystander effect exacerbates the problem, as passive peer observation can reinforce bullying behaviour and further isolate the victim.

The rise of online platforms and social media has ushered in a new era of verbal violence - cyber bullying. This form of abuse can be particularly insidious because it can reach victims anytime, anywhere, making it difficult to escape the relentless negativity. The anonymity and distance offered by online platforms often emboldens perpetrators, resulting in the spread of hateful messages, threats and harassment. Cyberbullying can take many forms, from spreading rumours and posting embarrassing photos to sending threatening messages and creating fake profiles to impersonate and harass the victim. The lack of face-to-face interaction can also contribute to the dehumanisation of the victim, making it easier for perpetrators to engage in cruel and hurtful behaviour. The psychological impact of cyberbullying can be profound, causing anxiety, depression, low self-esteem and even suicidal thoughts. The constant barrage of negative messages and the relentless nature of online harassment can overwhelm victims, creating a sense of helplessness and isolation. In addition, the permanence of online content means that the effects of cyberbullying can last long after the initial incident, leaving a lasting impact on the victim's self-image and reputation. The difficulty of identifying and holding perpetrators to account adds another layer of complexity to this already challenging issue.

The power dynamics in each context - family, workplace, school and online - play a crucial role in exacerbating verbal violence. In families, parental authority or sibling rivalry often creates power imbalances. In the workplace, hierarchical structures and the threat of job loss can silence victims. Schools foster social hierarchies that create vulnerability among students. Online, anonymity shields bullies and makes them less accountable. These power imbalances need to be addressed directly through policy changes, education and changes in social attitudes to create safer and more supportive environments. Addressing the underlying power imbalances is crucial to promoting effective prevention and intervention strategies. This requires tackling societal norms and structures that perpetuate these imbalances, and empowering victims to speak out and seek help without fear of reprisal. The creation of safe and accessible reporting mechanisms, coupled with effective consequences for perpetrators, is paramount. In addition, educating individuals about the different forms of verbal violence and its impact on victims is critical to cultivating a culture of empathy and respect.

The pervasiveness of verbal violence in these different contexts highlights the need for a multifaceted approach to address the issue. This includes comprehensive educational programmes to raise awareness, promote healthy communication skills and challenge societal norms that tolerate such behaviour. In addition, the establishment of clear policies and procedures for reporting and dealing with verbal violence in families, workplaces, schools and online platforms is crucial. Providing support and resources for victims, including counselling and therapy, is essential to help them heal and rebuild their lives. Creating a supportive environment where victims feel safe to come forward and seek help is critical to breaking the cycle of abuse. Finally, fostering a culture of accountability, where perpetrators are held accountable for their actions, is essential to prevent future instances of verbal violence and to ensure that all individuals are treated with respect and dignity. The collective efforts of individuals, organisations and policy makers are essential to create a world where verbal violence is no longer tolerated and where all people can feel safe and respected.

Defining profanity and its contextual meaning

Precisely defining profanity is a surprisingly complex task. What constitutes "profanity" varies dramatically across cultures, subcultures and even individual experiences. A word considered deeply offensive in one context may be commonplace or even humorous in another. This inherent contextual fluidity complicates any attempt at a rigid definition. Broadly speaking, profanity refers to language that is considered vulgar, obscene or disrespectful, often violating social norms and religious sensibilities. It includes words related to bodily functions, sexual acts and religious figures, as well as insults and epithets intended to demean individuals or groups.

However, the impact of a profane word is not only determined by the word itself. The context in which it's used - the speaker's intention, the audience's perception, the overall social setting and surrounding discourse - significantly shapes its meaning and effect. The casual use of a swear word among close friends might be considered acceptable, even bonding, while the same word directed at a stranger in a formal setting might be perceived as deeply offensive and aggressive. Tone of voice, body language and the relationship between speaker and listener all play a vital role in determining the overall impact.

Consider the difference between a frustrated individual muttering an expletive after stubbing his toe and a politician using the same word in a public speech to deliberately shock or offend his audience. The former is likely to be interpreted as a spontaneous expression of frustration, while the latter is a calculated act of provocation. The intention behind the profanity fundamentally alters its meaning and reception. Even in seemingly similar contexts, subtle shifts in intonation and body language can dramatically alter the perception of the profanity. A harsh, aggressive delivery turns a word into a weapon, whereas a softer, self-deprecating tone can make it almost humorous.

Cultural context has a significant impact on the interpretation and acceptance of profanity. What is considered profane in one culture may be perfectly acceptable or even commonplace in another.

Certain words or phrases considered highly offensive in Western societies may be used freely in other parts of the world, reflecting different cultural norms and values. A deep understanding of the cultural context is therefore crucial when assessing the use of profanity. In addition, sub-cultural groups often have their own unique linguistic norms and understandings of profanity. Within certain communities, slang and coded language may contain profane elements that are perfectly acceptable in that context, but would be incomprehensible or offensive to outsiders. This calls for a nuanced approach to the analysis of profanity, recognising the importance of linguistic diversity and its implications.

Furthermore, the evolution of language itself plays an important role in shaping perceptions of profanity. Words once considered highly offensive may lose their potency over time, becoming commonplace or even entering the mainstream lexicon. Conversely, new words or phrases may emerge as expressions of offence or rebellion, highlighting the constantly shifting nature of linguistic boundaries. The process of semantic change constantly redefines the boundaries of what constitutes profanity, rendering any static definition inadequate. Tracking these changes in language is crucial to understanding the dynamics of profanity's impact and its ever-shifting social connotations. The media, particularly through its exposure to diverse populations and diverse forms of expression, plays a powerful role in shaping these semantic shifts.

The use of profanity, whether intentional or unintentional, has a profound impact on communication. It can create barriers to understanding and prevent effective dialogue. Profane language often conveys aggression, disrespect and a lack of consideration for the listener. This can lead to strained relationships, damaged reputations and escalation of conflict. Conversely, profanity can be used strategically to enhance communication in certain settings. In informal settings, the shared use of profanity can build camaraderie and create a sense of shared identity among individuals. This requires a careful distinction between acceptable and abusive use of profanity.

The difference lies not only in the choice of words, but also in the intention behind their use and the context in which they're uttered.

Analysing the legal and ethical implications of profanity requires a delicate balance between freedom of speech and the need to protect individuals from harm. Legal frameworks in many countries regulate the use of profanity in public spaces, the media and the workplace, often with the aim of preventing offence and maintaining social order. However, the interpretation and enforcement of these laws can be complex and often controversial, particularly regarding the balance between restrictions and personal freedoms. Ethical considerations revolve around the potential impact of profanity on others, including the possibility of causing psychological distress, creating a hostile environment and perpetuating harmful stereotypes or prejudices. Again, the question of intent is crucial in determining ethical considerations - a spontaneous expression of frustration is very different from a deliberate attempt to cause harm or incite violence.

The use of profanity as a tool of verbal violence requires particular scrutiny. In this context, profanity goes beyond casual expression and becomes a weapon used to intentionally cause psychological harm, intimidate or humiliate others. The repetitive and targeted nature of such profanity, often combined with other forms of verbal abuse, can have devastating consequences for victims. The intent to cause harm is clear in such cases, making them ethically reprehensible and often legally actionable. Understanding the subtle ways in which profanity can be weaponised is crucial to developing effective strategies for prevention and intervention.

In conclusion, defining profanity and its contextual meaning is a multifaceted endeavour. It requires a deep understanding of cultural norms, social situations, linguistic dynamics and the intention of the speaker. While profanity can be used casually or even playfully, its use as a tool of verbal violence is a serious concern with significant legal and ethical implications. The ability to distinguish between acceptable and abusive profanity is a critical skill in promoting healthy communication and conflict resolution. This requires careful attention to the context of language, the intent behind it and the impact it has on others. The constant evolution of language and social norms requires constant critical analysis and reassessment of the boundaries of acceptable language.

Hate Speech Definitions Legalities and Implications

Hate speech, unlike abusive language, goes beyond mere offensiveness; it represents a deliberate attempt to dehumanise, marginalise and incite violence against individuals or groups on the basis of their perceived characteristics. While the line between offensive language and hate speech can sometimes be blurred, the crucial distinction lies in intent and effect. Offensive language may be used in frustration or anger, even playfully between intimates, but hate speech is inherently targeted and intended to cause harm through prejudice.

Defining hate speech accurately is a significant challenge, as its manifestations are diverse and context-dependent. Legal definitions vary from jurisdiction to jurisdiction and often focus on speech that incites violence, discrimination or hatred against a protected group. These protected groups typically include those defined by race, ethnicity, religion, national origin, sexual orientation, gender identity, disability or other characteristics. However, the lack of a universally agreed definition contributes to ongoing debates about its scope and application. Some argue that even seemingly innocuous speech can constitute hate speech if it contributes to a hostile environment or reinforces negative stereotypes.

Others argue that freedom of expression should protect even offensive speech unless it directly incites imminent violence. This tension between protecting freedom of expression and preventing harm lies at the heart of the legal and ethical complexities surrounding hate speech.

The impact of hate speech goes far beyond individual offence. It creates and perpetuates social divisions by normalising prejudice and discrimination. The constant bombardment of hateful rhetoric can have a chilling effect, silencing marginalised voices and discouraging open dialogue. Individuals targeted by hate speech often experience significant psychological harm, including anxiety, depression, and feelings of isolation and insecurity. The effects can be particularly acute for victims of sustained or systemic hate speech, where the constant barrage of negativity undermines self-esteem and contributes to a sense of powerlessness. Hate speech also creates an atmosphere of fear and intimidation, particularly within vulnerable communities. Members of marginalised groups may feel unsafe to express their identity or participate fully in society, limiting their opportunities and participation in public life.

There is an undeniable link between hate speech and discrimination. Hate speech doesn't just express prejudice, it actively cultivates and reinforces it. By repeatedly associating negative attributes with particular groups, hate speech creates and perpetuates harmful stereotypes and fosters a climate of hostility and intolerance.

This in turn can lead to tangible forms of discrimination, affecting access to employment, housing, education and other essential services. Victims of hate speech are more likely to experience discrimination in different aspects of their lives, creating a vicious circle in which hateful rhetoric fuels discriminatory practices.

Examining real-life examples highlights the devastating consequences of hate speech. The rise of online hate speech, facilitated by social media platforms, has created echo chambers where extremist views are amplified and normalised. Online harassment, targeted threats and the spread of misinformation can have a significant psychological and social impact on victims. Historical examples demonstrate the profound role of hate speech in inciting violence and genocide.

Even without direct calls to action, hate speech creates an environment of permissiveness that emboldens perpetrators of violence and undermines the protection of vulnerable populations.

Legal frameworks to combat hate speech vary widely from country to country. Some countries have robust laws criminalising hate speech, while others prioritise freedom of expression, even at the cost of allowing offensive or discriminatory speech. The challenge is to balance the need to protect vulnerable groups from harm with the fundamental right to freedom of expression. Many legal systems struggle to define hate speech with precision, leading to inconsistencies in enforcement and interpretation. In addition, the rapid development of communication technologies, particularly the rise of social media, poses new challenges for legal regulation. Existing laws often struggle to keep pace with the rapid spread of hate speech online, requiring constant adaptation and innovative approaches to regulation.

Beyond the legal framework, many social initiatives are working to counter hate speech. Education plays a crucial role in challenging prejudice and promoting the critical thinking skills needed to recognise and resist hate speech. Media literacy programmes equip individuals with the tools to critically evaluate information and identify biased or manipulative rhetoric. Community-based organisations often lead anti-hate campaigns, promote dialogue and provide support to victims of hate speech.

These efforts aim to create a more inclusive and tolerant society by tackling hate speech at its source and supporting those affected by it. In addition, social media platforms themselves are increasingly recognising their role in the spread of hate speech and are taking steps to remove harmful content and prevent its dissemination. However, these efforts are often criticised for their inconsistency and lack of transparency, highlighting the complexity of moderating online speech while upholding freedom of expression.

Tackling hate speech requires a multi-pronged approach that includes legal, social and technological measures. Robust legal frameworks are essential to deter hate speech and provide redress for victims. But laws alone are not enough; education, community engagement and media literacy are equally important to promote tolerance and challenge prejudice. Social media platforms have a responsibility to actively combat the spread of hate speech on their platforms, but transparency and accountability are crucial to ensure that their efforts are effective and do not violate freedom of expression.

The fight against hate speech is an ongoing challenge that requires constant adaptation and cooperation between governments, civil society organisations and technology companies. It is a fight for a fairer and more equal society where all people feel safe and respected, free from the corrosive effects of prejudice and hate. The task ahead requires a sustained and coordinated effort to combat not only the language of hate, but also the underlying social and political factors that give rise to it. The ultimate goal is not just to silence hate speech, but to change the social conditions that allow it to flourish. This requires a change in social attitudes, a renewed commitment to empathy and understanding, and constant vigilance against the insidious erosion of tolerance and respect.

The Psychology of Hate Speech Motivations and Mechanisms

The psychology of hate speech is a complex area that intertwines individual motivations with broader societal factors. Understanding why individuals engage in such harmful communication requires exploring the intricate interplay of cognitive biases, emotional responses and social dynamics. Fear, a primal human emotion, often plays an important role. Individuals may express hate speech as a way of coping with anxieties about perceived threats to their social group, economic security or cultural identity. The process of "othering", in which individuals or groups are categorised as different and inferior, often stems from fear and insecurity and leads to derogatory language and discriminatory actions. This is particularly evident in times of social and political upheaval, when scapegoating and prejudice become more prevalent. For example, during economic downturns, immigrants or minority groups are often targeted as the cause of societal problems, fuelled by fear and resentment.

Insecurity, which is closely linked to fear, can also fuel the production of hate speech. Individuals who lack self-esteem or feel threatened by their social status may seek to bolster their self-image by denigrating others. Hate speech can become a tool for asserting dominance or belonging within a particular group, creating a sense of community based on shared prejudices. This dynamic is evident in online echo chambers, where individuals reinforce each other's prejudices and engage in increasingly extreme forms of hate speech, creating a spiral of negativity and hostility. For example, online forums dedicated to specific conspiracy theories often contain hateful rhetoric targeting individuals or groups perceived as enemies of the movement. The sense of belonging and common purpose outweighs any concern about the harmful nature of the language used.

The desire for social dominance is another powerful motivator. Hate speech can be a way of asserting power and control over marginalised groups, reinforcing existing hierarchies and social inequalities. Historically, hate speech has been used strategically to justify oppression and violence against minority groups, stripping them of their dignity and humanity. The dehumanising language used in propaganda campaigns throughout history underlines this point and creates a climate of acceptance for discrimination and violence.

Examples include the anti-Semitic rhetoric that preceded the Holocaust and the racist language used to justify slavery and segregation. These historical examples demonstrate that hate speech is not simply a by-product of prejudice, but an active tool for maintaining and exacerbating power imbalances.

Beyond individual motivations, the cognitive processes involved in the production and reception of hate speech are crucial. Cognitive biases such as confirmation bias (seeking information that confirms pre-existing beliefs) and in-group bias (favouring members of one's own group over others) play an important role in shaping individuals' perceptions and attitudes. These biases can lead people to interpret information selectively, reinforcing their prejudiced views and making them more vulnerable to hate speech. For example, individuals with pre-existing prejudices against immigrants may interpret ambiguous news stories in a way that confirms their negative stereotypes, potentially leading to the spread of hate speech.

Social learning theory sheds further light on the spread of hate speech. Individuals often learn prejudiced attitudes and behaviours by observing others, particularly those they admire or identify with. This process of social learning is particularly prevalent in families, peer groups and communities where discriminatory language and beliefs are normalised. Children who grow up hearing hateful remarks from family members or peers are more likely to adopt these attitudes and behaviours themselves. Similarly, exposure to hate speech in the media, whether through online platforms or traditional media, can normalise such language and increase the likelihood of its use and dissemination.

Equally important are the emotional mechanisms involved in hate speech. Anger, resentment and frustration often fuel the expression of prejudiced language. Individuals may use hate speech to vent their emotions, even if they do not genuinely believe in the harmful stereotypes they are expressing. However, the act of expressing hateful rhetoric can itself reinforce negative feelings and attitudes, creating a vicious cycle. The emotional satisfaction derived from expressing anger or resentment through hate speech can lead to repeated engagement in such behaviour, reinforcing existing prejudices.

In addition, the interplay between individual psychology and social context has a significant impact on the creation and dissemination of hate speech. Social norms and cultural values can either condone or condemn hate speech, creating different environments in which prejudiced speech is more or less likely to flourish. In societies where discrimination and prejudice are widely accepted, hate speech is more likely to be tolerated, even normalised, while societies with strong anti-discrimination norms may actively suppress the expression of hateful rhetoric. The role of social media in this context is particularly striking. Online platforms, through algorithms and echo chambers, can amplify hate speech and create conditions for its rapid spread, while also creating spaces where individuals can feel isolated and emboldened in their prejudiced views.

The reception of hate speech also involves a complex interplay of psychological factors. Victims of hate speech often experience significant emotional distress, including feelings of anger, fear, shame and humiliation. These experiences can lead to psychological damage, including anxiety, depression and post-traumatic stress disorder. Repeated exposure to hate speech can contribute to feelings of powerlessness and vulnerability, which negatively affect self-esteem and mental well-being. The psychological consequences of hate speech can have long-lasting effects on individuals' lives, affecting their relationships, social participation and even their physical health.

Defining bullying and its verbal manifestations

Defining bullying requires a nuanced understanding that goes beyond simple aggression. While all forms of bullying involve an imbalance of power, the defining characteristic is the repetitive and targeted nature of the behaviour. It's not a single incident of anger or disagreement, but a pattern of actions designed to harm, intimidate or control another individual. This deliberate and sustained negativity distinguishes bullying from other forms of conflict or disagreement. For example, a single insult may be hurtful, but it doesn't necessarily constitute bullying unless it's part of a larger pattern of harassment. The key is a consistent and deliberate attempt to cause psychological or emotional distress.

Verbal bullying is particularly insidious because it often leaves no visible mark. Unlike physical bullying, the wounds inflicted are less tangible, but no less real. The cumulative effect of constant criticism, insults, threats and humiliation can have a profound and long-lasting impact on the victim's self-esteem, mental health and general wellbeing. The perpetrator's words become a constant source of anxiety and stress, shaping the victim's perception of themselves and their place in the world. This constant barrage can lead to feelings of isolation, helplessness and depression, affecting their academic or professional performance and even leading to physical health problems.

Consider the seemingly harmless but deeply damaging effects of persistent teasing. A child who is repeatedly teased about his or her appearance, or a colleague who is constantly ridiculed for his or her ideas, may not experience a single catastrophic event, but rather a slow erosion of self-confidence and self-esteem. The cumulative weight of these micro-aggressions, these seemingly small acts of verbal aggression, can be devastating. These repeated negative interactions create a toxic environment that undermines the target's sense of belonging and safety. The persistent nature of these verbal attacks is key to defining them as bullying, as repeated exposure to negativity creates a cumulative harm that goes far beyond a single incident.

A critical element in understanding verbal bullying is the distinction between direct and indirect forms. Direct verbal bullying involves overt acts of aggression such as name-calling, insults, threats and intimidation. This is the most recognisable form of verbal bullying and is characterised by the perpetrator directly confronting the victim with hostile language. Examples include shouting insults across a classroom, spreading rumours about someone's private life to damage their reputation, or making persistent threats to cause harm. The direct nature of these attacks often makes it easier to identify and report bullying behaviour, but the damage caused is no less damaging than more insidious forms of verbal bullying.

Indirect verbal bullying is more subtle and insidious, often done behind the victim's back. It can include spreading malicious rumours, gossip, social exclusion and cyberbullying. This type of bullying is more difficult to identify and deal with because it often lacks the direct confrontation of face-to-face bullying. The victim may not always know the source of the negativity, leading to feelings of confusion, anxiety and isolation.

The lack of direct confrontation makes it difficult to confront the perpetrators directly, creating more challenges in dealing with this form of bullying.

Cyber bullying, a particularly insidious form of indirect verbal bullying, uses the anonymity and reach of online platforms to cause harm. The use of social media, messaging apps and online forums allows perpetrators to target victims relentlessly, often without facing immediate consequences. The persistent nature of online harassment, where hateful messages and images can be shared and re-shared indefinitely, exacerbates the emotional distress experienced by the victim. The ability to reach the victim anytime, anywhere increases the sense of being relentlessly pursued and harassed. This aspect of relentless pursuit and lack of privacy creates a unique and particularly harmful dimension to cyberbullying. The ease of creating fake profiles and hiding one's identity also adds to the difficulties in addressing and stopping this form of bullying.

Social exclusion, another form of indirect verbal bullying, involves the deliberate exclusion of an individual from social groups or activities. This can range from being ignored or ostracised to being systematically excluded from conversations or social events. While this may not involve direct verbal aggression, the act of deliberate exclusion can be just as damaging, creating a sense of isolation and rejection. This exclusion can undermine an individual's sense of belonging, social identity and self-worth. It can be extremely difficult to address as there are often no clear indicators of targeted harassment and it involves subtle but equally damaging social dynamics.

The impact of power imbalances further complicates the dynamics of bullying. In schools, for example, older or larger pupils often have a power advantage over younger or smaller ones, making them more likely to engage in bullying behaviour. Similarly, in the workplace, supervisors or senior colleagues may use their authority to intimidate or harass junior staff. This power differential allows perpetrators to use their position to harm those with less influence. The victim's ability to defend themselves is often compromised by the power imbalance. This often leaves them feeling powerless, helpless and unable to seek help for fear of further repercussions.

The power imbalance inherent in bullying is rarely explicit; it operates through subtle cues and actions. It's in the knowing looks, the shared laughter at someone's expense, the knowing silence that allows the bullying to continue unchecked. It's the unspoken understanding within a group, the social dynamics that prevent intervention and allow the bullying to continue. This creates a climate of fear and intimidation that makes it difficult for victims to report the bullying, seek support or even acknowledge that they're being bullied. It's the unspoken rules and norms within these systems that create the conditions for bullying and make it so difficult to eliminate.

The long-term consequences of verbal bullying can be severe, affecting victims' mental and physical health, academic and work performance, and relationships. The emotional scars of persistent verbal abuse can last well into adulthood, affecting self-esteem, confidence and the ability to form healthy relationships. Victims may experience anxiety, depression, post-traumatic stress disorder and other mental health problems. In severe cases, verbal bullying can contribute to self-harm or suicidal thoughts. The impact on academic or work performance can be significant, with victims having difficulty concentrating, participating effectively in class or work, or maintaining healthy work relationships.

In addition, the impact on their personal lives can be equally debilitating, as they may struggle to trust others or form meaningful relationships, leading to social isolation and a reduced quality of life.

Tackling verbal bullying requires a multi-pronged approach involving victims, perpetrators, bystanders and the wider community. Victims need support and resources to cope with the effects of bullying, including counselling, support groups and educational programmes. Bullies need to understand the impact of their actions and develop strategies to change their behaviour.

This may include interventions such as anger management programmes, conflict resolution training or restorative justice practices. Bystanders have an important role to play in preventing bullying.

Encouraging bystanders to intervene when they witness bullying behaviour, even in small ways, can significantly reduce the occurrence of bullying incidents. Ultimately, changing the overall environment to promote a culture of respect, empathy and inclusion will help prevent future incidents and support the recovery of those affected. This requires a community-wide effort involving schools, workplaces, families and the wider community, with a consistent message of zero tolerance towards bullying behaviour. Such a comprehensive approach can create a safer and more supportive environment for all.

The Role of Power Dynamics in Verbal Bullying

The insidious nature of verbal bullying is often reinforced by underlying power dynamics. These aren't always overt; they can be subtle and complex, operating through social structures, hierarchical relationships and even unspoken group norms.

Understanding these power dynamics is crucial to effectively addressing and mitigating verbal bullying. One of the most common manifestations of this power imbalance is age and size differences, which are particularly prevalent in school settings. Older children often have a significant power advantage over younger or smaller peers due to their physical maturity and perceived social status. This power differential allows them to engage in bullying behaviour with less fear of immediate consequences. A larger pupil can easily intimidate a smaller one, using verbal threats to maintain control and dominance. The intimidation is not only physical but also psychological; the perceived threat of physical violence allows verbal abuse to flourish.

The workplace is a different but equally important arena where power dynamics can fuel verbal bullying. Supervisors, managers and senior colleagues often wield considerable authority over their subordinates. This inherent power imbalance creates an environment where verbal aggression, however subtle, can go unchecked. A manager can consistently criticise an employee's work, use condescending language and undermine their confidence without fear of significant reprisal. The employee, who relies on the job for income and status, may be reluctant to report the abuse for fear of professional repercussions. This silent acceptance of verbal abuse reinforces the power dynamic and allows the bullying to continue.

Social status is also an important factor. In any social group, individuals with higher social status - whether through popularity, athletic prowess or academic achievement - can use their position to exert control and influence over others. This power can manifest itself in various forms of verbal bullying, from subtle exclusion to direct insults and humiliation. Popular cliques can use their social power to exclude or ridicule those perceived as 'outsiders', creating a hostile environment in which victims feel isolated and helpless.

Victims' fear of social isolation often prevents them from reporting or resisting this form of bullying, further consolidating the power and dominance of the bullies.

Beyond individual relationships, group dynamics play an important role in exacerbating verbal bullying. The collective power of a group can be much greater than the sum of its individual members. When a group is involved in bullying, the victim faces not just one aggressor, but a multitude. This multiplies the psychological impact and limits the victim's ability to defend themselves or seek support. The power of the group creates a sense of anonymity and diffusion of responsibility for individual members. Even if a member feels uncomfortable with the bullying, they may be reluctant to speak out for fear of being excluded from the group or facing retaliation. The perceived risk of challenging the group's behaviour is often greater than the risk of remaining silent, allowing the bullying behaviour to escalate and become more entrenched.

Online interactions provide a unique and increasingly common context for verbal bullying influenced by power dynamics. The anonymity afforded by online platforms can embolden perpetrators, giving them a sense of impunity that allows them to engage in more aggressive and unrestrained verbal attacks. Trolls and cyber bullies often exploit this anonymity to target individuals, often with little fear of immediate consequences. This anonymity creates a false sense of security for bullies, allowing them to express hateful views and engage in aggressive behaviour. This power dynamic is further complicated by the relatively unregulated nature of online environments, where effective intervention is less readily available and easier to circumvent.

The power dynamics at play in cyber bullying are exacerbated by the sheer reach and speed of online communication. Harmful messages and images can be shared instantly with a wide audience, causing significant damage to the victim's reputation and emotional well-being. The ability to reach the victim anytime, anywhere reinforces the sense of being relentlessly stalked and harassed, making it difficult to escape the negativity. This constant exposure to online aggression significantly increases the intensity of the bullying and its impact on the victim's mental health. The relentless and pervasive nature of online harassment presents unique challenges in addressing and mitigating the harm caused.

In addition, the power dynamics of online bullying often manifest themselves in the creation of online echo chambers where like-minded individuals reinforce each other's biases and prejudices. This group reinforcement of negative opinions and behaviours can embolden bullies and amplify the harmful effects of their actions. The victim is not dealing with just one bully, but with a collective expressing similar negative feelings. This makes it harder to find support and counter the negative narratives. Breaking through these echo chambers and challenging the prevailing group dynamics is crucial to tackling this form of verbal bullying.

The impact of power dynamics on verbal bullying goes beyond the immediate consequences. The long-term effects on victims can be profound and far-reaching. The constant barrage of verbal abuse can undermine self-esteem, leading to anxiety, depression and even physical health problems. The victim's sense of self-worth is diminished, affecting their academic and professional performance, as well as their ability to form healthy relationships.

The constant stress and anxiety caused by ongoing bullying negatively affects every aspect of life. Effectively addressing verbal bullying requires a nuanced understanding of these power dynamics. Interventions need to take into account the social structures, hierarchies and group dynamics that contribute to the perpetuation of such behaviour.

Strategies need to focus not only on the immediate actions of bullies, but also on the underlying power structures that enable and reinforce their behaviour. Creating a culture of respect, empathy and inclusivity - one in which individuals feel empowered to speak up and challenge bullying behaviour - is paramount to creating a safer and more equitable environment for all. This requires a concerted effort by individuals, institutions and communities to challenge the power imbalances that fuel verbal violence and to create spaces where all voices are heard and valued. Only then can we hope to effectively address and mitigate the harmful effects of verbal bullying and promote healthier patterns of communication.

Addressing the power dynamics involved in verbal bullying requires a multi-faceted approach. It requires not only dealing with individual perpetrators, but also addressing the systemic issues that allow bullying to flourish. Educational initiatives are crucial in promoting critical awareness of power dynamics within social groups. These programmes should equip individuals with the skills to identify and challenge power imbalances, promote empathy and encourage bystander intervention. Workplace policies must actively address verbal harassment by providing clear channels for reporting and robust procedures for investigation and disciplinary action. By holding institutions accountable for maintaining respectful and inclusive environments, the power dynamics that enable verbal abuse can be systematically challenged.

Victim empowerment is also essential. Support networks, counselling services and legal aid should be readily available to those who have experienced verbal harassment. These support systems are essential for victims to regain control of their lives and to heal from the psychological trauma of verbal abuse. The aim is not just to deal with individual cases, but to create a system where victims feel safe and empowered to come forward without fear of reprisal or further victimisation.

Finally, broader societal changes are needed to challenge the normalisation of verbal aggression. Promoting positive communication skills and conflict resolution techniques from an early age is essential. Challenging negative stereotypes and prejudices that can fuel verbal aggression is also crucial. These broader cultural changes require a collective effort in which families, schools, workplaces and communities work together to create a culture of respect and empathy in which verbal bullying is actively challenged and prevented. Only through such a comprehensive approach can we truly hope to overcome the power imbalances that underpin verbal bullying and create a society where everyone feels safe and respected.

The psychological impact of bullying on victims

The insidious nature of verbal bullying goes far beyond the immediate sting of hurtful words. Its psychological impact on victims is profound and long-lasting, leaving scars that can significantly affect their self-esteem, mental health and social relationships for years to come. The constant barrage of negativity, the constant undermining of their confidence and the feeling of powerlessness can be devastating.

One of the most immediate and significant effects is the erosion of self-esteem. Repeated verbal attacks, whether direct insults or subtle put-downs, chip away at a victim's sense of self-worth. They begin to internalise the negative messages and believe the hurtful things that are said about them. This internalised negativity can manifest as self-doubt, feelings of inadequacy and a diminished sense of personal worth. The victim may begin to question their abilities, their intelligence and even their worthiness of love and acceptance.

This can lead to a cycle of self-criticism and negative self-perception, making it difficult to achieve goals and maintain positive relationships.

The impact on mental health can be equally serious. Victims of verbal bullying have a significantly higher risk of developing anxiety, depression and other mental health problems.

The constant stress and anxiety associated with bullying can lead to sleep problems, changes in appetite and physical symptoms such as headaches and stomachaches. The feeling of constant threat, of never knowing when the next attack will come, creates a state of chronic hypervigilance that takes its toll on both mind and body. The constant fear can make it difficult to concentrate on daily activities, both academic and personal, as their attention is constantly guarded against the possibility of the next instance of abuse.

The social impact of bullying is also significant. Victims often withdraw from social interactions for fear of further humiliation or rejection. Their trust in others is broken, making it difficult to form healthy and meaningful relationships. They may become isolated and lonely, further exacerbating their mental health problems. Fear of social interaction can lead to avoidance behaviour, increasing their isolation and exacerbating the effects of bullying. This social isolation is often unintentional, but is a direct consequence of the negative effects of verbal abuse.

Academic performance often suffers as a result of verbal bullying. The constant stress and anxiety affects concentration and learning, leading to lower academic achievement. Victims may struggle to concentrate in class, find it difficult to participate in discussions and experience a general lack of motivation. The psychological toll of bullying makes even the simplest academic tasks seem overwhelming, leading to a decline in performance and a sense of failure. This academic decline can create a vicious cycle in which poor performance further lowers self-esteem and increases vulnerability to further bullying.

The long-term effects of verbal bullying can be far-reaching and deep-seated. People who are verbally bullied in childhood or adolescence may carry these emotional scars into adulthood. They may struggle to maintain healthy relationships, experience difficulties at work and have an increased sensitivity to criticism. The lasting impact on their self-esteem can affect their career choices, social life and even romantic relationships. The effects can extend into every facet of their adult lives, demonstrating the severity of the psychological toll of verbal bullying.

The coping mechanisms used by victims of verbal bullying vary widely. Some may try to ignore the bullying in the hope that it will eventually stop. Others may try to retaliate by engaging in verbal conflict with their bullies. Some may withdraw from social situations and isolate themselves to avoid further harm. Still others may seek support from friends, family or mental health professionals. The effectiveness of these coping mechanisms depends largely on the severity and persistence of the bullying, as well as the individual's resilience and access to support systems. It's important to recognise that not all coping mechanisms are equally effective. The method a victim chooses will depend on individual resilience, available support systems and personal understanding of the situation.

The effectiveness of ignoring bullying varies greatly. While some cases of low-level bullying may stop when ignored, more persistent or severe forms of verbal abuse are unlikely to stop unless addressed more directly. Retaliation, often used out of frustration and a desire to assert dominance, can escalate the situation and lead to further conflict, exacerbating the psychological impact. Withdrawal, while offering a temporary reprieve from the immediate negativity, can lead to social isolation and a decline in mental wellbeing. Seeking support is therefore often the most effective long-term strategy.

Seeking professional help is often crucial for victims of verbal bullying. Therapy can provide a safe space to process feelings, develop coping strategies and build resilience. Mental health professionals can help people understand the impact of bullying on their mental health and develop healthy ways of dealing with their feelings. They can also work with individuals to develop effective communication skills and strategies for managing future conflict, and to build confidence and assertiveness. In addition, support groups provide opportunities for connection, shared experiences and collective healing, highlighting the importance of a wider support system.

Early intervention is crucial in mitigating the long-term effects of verbal bullying. Addressing the issue promptly can prevent the development of serious mental health problems and promote healthy emotional development. Schools, workplaces and other institutions have a crucial role to play in identifying and addressing verbal bullying. This can include implementing anti-bullying programmes, providing support services for victims and holding perpetrators accountable for their actions. The role of bystanders is also important; encouraging them to intervene and support victims can significantly reduce the severity and duration of bullying.

The role of education in preventing verbal abuse

Education plays a key role in preventing verbal violence, as it is a fundamental pillar in cultivating a society where respectful communication thrives. It's not just about imparting knowledge; it's about fostering empathy, understanding and the development of critical social-emotional skills that enable individuals to manage conflict constructively. This preventive approach starts early and focuses on shaping children's attitudes towards communication and conflict resolution before harmful patterns become entrenched. The home and school environments are particularly important in the formative years as they provide the initial framework for healthy interactions.

Parents, as primary caregivers, have an important responsibility to model respectful communication. Children learn by observing the dynamics within their families. If parents frequently engage in verbal aggression, insults or belittling remarks towards each other or their children, children are likely to internalise these behaviours as acceptable norms. Conversely, a family environment characterised by open dialogue, empathy and respectful disagreement teaches children invaluable skills for managing conflict constructively. Active listening, respectful expression of feelings and compromise become integral parts of their communication repertoire.

Parents can actively promote this by engaging in family discussions where differing opinions are valued, conflicts are addressed calmly and solutions are sought together. They can also encourage their children to express their feelings, model effective communication strategies and provide opportunities to practice respectful disagreement.

The school environment extends parental influence by providing a wider social context in which children interact with peers and authority figures. Educators have a crucial role in shaping children's understanding of healthy communication and conflict resolution.

Integrating social-emotional learning (SEL) programmes into the curriculum is critical. SEL programmes focus on developing children's self-awareness, self-management, social awareness, relationship skills and responsible decision-making. These skills are essential for preventing verbal violence because they enable children to understand their own emotions, manage their impulses, empathise with others, build positive relationships and make decisions that minimise harm.

Implementing effective SEL programmes requires careful planning, well-trained educators and ongoing evaluation of the impact on the school community.

Effective SEL programmes are not simply add-ons to the curriculum; they are woven into the fabric of the educational experience. They are woven into subjects ranging from literature and history to science and mathematics. For example, analysing the interactions of characters in literature can help students to identify different communication styles and their impact on relationships. Discussions about historical conflicts can highlight the consequences of aggressive communication and the importance of finding peaceful solutions. Even in mathematics, cooperative problem-solving activities can promote teamwork and respectful collaboration. Integrating SEL across subject areas reinforces the message that respectful communication is an important life skill that applies to all aspects of life.

Beyond structured SEL programmes, educators can actively promote a climate of respect and understanding in the classroom. Creating a classroom culture where every student feels safe to express their ideas, ask questions and participate in discussions is paramount. This requires educators to actively listen to students, acknowledge their perspectives, and manage conflict fairly and calmly. Effective classroom management techniques, including positive reinforcement, clear expectations and consistent consequences for aggressive behaviour, are essential to creating a positive learning environment. Teachers should also model respectful communication, demonstrating empathy, understanding and conflict resolution skills in their interactions with pupils and colleagues. This modelling is crucial because children learn as much from observing adult behaviour as from explicit instruction.

It's also important to include conflict resolution skills in the curriculum. Teaching children techniques for managing disagreements peacefully, including active listening, negotiation and compromise, helps them develop the skills to resolve conflicts without resorting to verbal violence. Role-playing and group discussions can provide opportunities for pupils to practice these skills in a safe and supportive environment.

It is also important to promote empathy and perspective taking. Activities that help students understand the feelings and experiences of others can foster compassion and reduce the likelihood of harmful verbal interactions. These activities can range from simple exercises in imagining someone else's perspective to more complex projects involving community service or collaboration on social issues.

Creating a positive school climate is not the sole responsibility of educators; it requires the combined efforts of students, parents and the wider school community. School policies that explicitly prohibit verbal violence and bullying are essential. These policies should clearly define unacceptable behaviour and outline the consequences for violating them. Equally important is the establishment of clear reporting mechanisms that allow students to report incidents of verbal abuse without fear of retaliation.

Investigating all reported incidents thoroughly and fairly is critical to maintaining trust and ensuring accountability.

However, prevention efforts extend beyond the classroom and the family. Community-based programmes that promote conflict resolution and positive communication skills can further enhance the effectiveness of school-based initiatives. These programmes may involve partnerships between schools, community organisations and local businesses to provide workshops, support groups or mentoring opportunities for young people. Such collaborative efforts can create a more cohesive and supportive network that reinforces the messages of respect and understanding taught in schools and at home.

The role of technology in facilitating communication cannot be ignored. The internet and social media have become ubiquitous platforms for interaction, creating new avenues for verbal abuse.

Educating children and young adults about responsible online behaviour is crucial to mitigating the harmful effects of cyber bullying and online harassment. This includes teaching them about the potential consequences of their online actions, the importance of empathy and respect in online communication, and the resources available for support if they experience or witness cyber bullying. Schools and families should work together to set guidelines for online behaviour and ensure that children understand the seriousness of cyber bullying and the need to report such incidents.

Community-based initiatives to promote respect and understanding

Community-based initiatives are crucial in extending the reach of efforts to prevent verbal violence beyond the confines of schools and families. These initiatives create a supportive network that reinforces the values of respect, empathy and understanding and promotes a culture of positive communication within the wider community. Successful programmes often involve collaboration between schools, community organisations, local businesses, law enforcement agencies and faith-based institutions, drawing on the strengths and resources of each partner to achieve a common goal.

One effective strategy is to develop community-wide campaigns to raise awareness of verbal violence. These campaigns can use various media channels, including posters, flyers, social media and public service announcements, to disseminate information about the impact of verbal abuse, the importance of respectful communication and the resources available to victims. Campaigns should use clear, concise messages and avoid jargon or overly technical language to ensure widespread understanding and accessibility. They should also highlight the positive aspects of respectful communication and the benefits of creating a more compassionate community.

In addition to awareness campaigns, community-based initiatives often focus on providing direct services to victims of verbal abuse. Support groups, counselling services and mentoring programmes provide safe spaces where individuals can share their experiences, receive emotional support and learn coping mechanisms. These services are particularly important for vulnerable groups, such as children, the elderly and people with disabilities, who may be at increased risk of verbal abuse and may face additional barriers to accessing support. The establishment of confidential and accessible helplines to provide immediate support and advice to those in need is also essential. These support services should be staffed by trained professionals who understand the dynamics of verbal violence and are able to provide trauma-informed care.

Another important aspect of community-based initiatives is the establishment of conflict resolution programmes. These programmes provide training and workshops in various conflict resolution techniques, including mediation, negotiation and restorative justice.

Participants learn practical skills to manage disagreements peacefully, avoid escalation and find mutually acceptable solutions. These workshops should include interactive exercises, role-plays and group discussions to allow participants to practice their skills in a safe and supportive environment. Programmes should cater for different age groups and cultural backgrounds to ensure inclusivity and relevance to the specific needs of the community. Community centres, libraries and faith-based organisations can be ideal locations for these programmes, as they provide easily accessible venues for local residents.

Community-based initiatives can also play an important role in fostering positive relationships within the community. By organising events that bring people together, such as community festivals, sporting events and volunteering projects, these initiatives create opportunities for interaction and collaboration. These events can help break down barriers, foster a sense of community and create a more inclusive environment where everyone feels valued and respected. Such collaborative efforts can build social cohesion and strengthen community ties, reducing the isolation that often accompanies experiences of verbal violence. Focusing on collaborative, positive activities helps to shift the community's focus away from negativity and towards positive interactions.

The involvement of local organisations and community leaders is crucial to the success of community-based initiatives. These individuals can act as role models, advocates and mentors, demonstrating the importance of respect and understanding in their interactions with others. They can also help mobilise community resources, secure funding and recruit volunteers. Community leaders can act as liaisons between different groups and organisations, facilitating cooperation and communication. Their active involvement creates a sense of ownership and commitment within the community, increasing the sustainability and long-term impact of initiatives.

In addition, partnerships with local businesses can provide valuable resources and support for community-based initiatives. Companies can provide space for workshops, donate funds, provide in-kind services, or offer internships or volunteer opportunities. This mutually beneficial relationship demonstrates corporate social responsibility and contributes to a positive community image, while supporting critical efforts to prevent verbal violence.

These partnerships can also help raise awareness in the wider community, extending the reach and impact of initiatives.The use of restorative justice practices is also gaining traction as a valuable strategy in community-based initiatives. Restorative justice focuses on repairing the harm caused by conflict and promoting healing for both victims and offenders. It involves bringing together all parties to a conflict to discuss the impact of their actions and find ways to make amends. This approach can be particularly effective in cases of verbal violence, as it allows victims to share their experiences, perpetrators to take responsibility for their actions, and the community to play a role in repairing the harm caused. The collective involvement of the community strengthens social ties and fosters a greater sense of shared responsibility in preventing future incidents. Careful implementation and training are essential to ensure that restorative justice is effective and doesn't inadvertently re-traumatise victims.

Ongoing evaluation and assessment are essential to ensure the long-term effectiveness of community-based initiatives. Collecting data on the reach and impact of programmes, feedback from participants and input from the community is essential for making improvements and adapting to changing needs. Regular evaluation allows for course correction and provides evidence to secure continued funding and community support. It demonstrates accountability and a commitment to continuous improvement, thereby enhancing the credibility and sustainability of the initiative. The data collected can inform future planning and ensure that resources are allocated efficiently to address specific challenges and needs within the community.

The role of media and technology in shaping communication norms

The pervasiveness of media and technology in modern life has profoundly reshaped communication norms, influencing not only how we interact, but also the nature of conflict and the prevalence of verbal violence. Idealised portrayals of conflict in film and television, which often glorify aggression and minimise the consequences of hurtful words, contribute to a distorted understanding of healthy communication. These narratives often portray verbal sparring as witty banter or a sign of strength, subtly normalising aggressive communication styles and undermining the importance of respectful interaction. Characters who engage in verbal abuse often experience little to no negative repercussions, further reinforcing the acceptance of such behaviour.

The lack of consequences in fictional narratives can desensitise audiences to the harmful effects of verbal abuse, blurring the lines between playful teasing and outright aggression. This normalisation extends beyond entertainment media; news coverage can also inadvertently shape public perceptions of acceptable communication styles. Sensationalised coverage of conflict, focusing on inflammatory statements rather than underlying issues, contributes to a climate in which aggressive communication is not only tolerated, but in some cases celebrated.

In addition, the anonymity and perceived distance offered by online platforms exacerbates the problem. The lack of face-to-face interaction can alienate individuals, making them more likely to engage in aggressive communication online than in person. The ease with which hateful comments can be posted and shared, often with little fear of immediate repercussions, fuels the spread of online harassment and cyber bullying. The anonymity of the Internet, coupled with the potential for rapid dissemination of harmful content, creates an environment ripe for verbal violence. The echo chambers of social media, where individuals interact primarily with like-minded people, can amplify existing prejudices and reinforce extreme viewpoints, fostering a climate of intolerance and hostility. This online environment allows for the spread of hate speech and the perpetuation of harmful stereotypes, directly impacting offline interactions and contributing to a culture where respectful dialogue is increasingly difficult.

The rise of social media has fundamentally changed patterns of communication, creating both opportunities and challenges for conflict resolution. While social media platforms can facilitate connection and foster community, they can also become breeding grounds for conflict. The immediacy of online communication, combined with the potential for misinterpretation due to the lack of non-verbal cues, can lead to misunderstandings and escalating disagreements. The lack of immediate feedback and the potential for comments to be taken out of context increase the risk of miscommunication and conflict. In addition, the permanence of online communication presents a unique challenge. Once a comment or message is posted, it remains accessible and can cause lasting damage to reputations and relationships. The difficulty of effectively mediating or resolving online conflict adds another layer of complexity, requiring specialised strategies and interventions.

The responsibility of media and technology companies to promote respectful communication and combat online harassment cannot be overstated. The media has an important role to play in shaping public attitudes towards verbal violence. By consciously choosing to portray respectful communication styles in their programming and news coverage, and by highlighting the negative consequences of verbal abuse, they can contribute to a culture change. It is crucial for media organisations to implement robust editorial guidelines that ensure responsible reporting of conflict, focusing on facts and avoiding sensationalisation of inflammatory language. Similarly, technology companies have an important responsibility to combat online harassment. The development and implementation of effective content moderation policies, coupled with improved mechanisms for reporting and addressing online abuse, are critical to creating safer online environments. Platforms should invest in technology and policies to quickly and effectively identify and remove hate speech, cyber bullying and other forms of online harassment.

Education and training of users on responsible online behaviour is also essential. Initiatives to raise awareness of the harmful effects of online harassment and to promote respectful communication online are essential. Such initiatives should focus on developing digital literacy skills, teaching users to recognise and avoid online abuse, and empowering them to take action against harmful content. By working together with educators, community organisations and mental health professionals

technology companies can develop comprehensive strategies that address the root causes of online harassment and promote a culture of respect and responsibility. This collaborative approach is critical, as individual efforts by media or technology companies alone are unlikely to have the desired impact.

Beyond policy changes, a broader societal shift in attitudes towards communication is needed. The normalisation of aggressive communication, both online and offline, needs to be challenged.

Encouraging empathetic communication skills and promoting understanding of different points of view is essential. This includes promoting media literacy education at all levels, from primary school to higher education, to enable individuals to critically analyse media representations and recognise the potential harm of aggressive communication. Education should include not only understanding the impact of harmful language, but also learning skills for effective communication, conflict resolution and the development of empathy.

Integrating respectful communication strategies into different settings - schools, workplaces and community organisations - is equally important. Establishing clear guidelines for acceptable behaviour, providing training in conflict resolution and creating safe spaces for dialogue are crucial to fostering a culture of respect. These efforts need to be complemented by public awareness campaigns that emphasise the importance of respectful communication and highlight the negative impact of verbal violence on individuals and society. Such campaigns should not only provide information but also inspire action, encouraging individuals to challenge aggressive communication styles and promote constructive dialogue.

Enforcement - Delays in investigation and prosecution

The crippling effect of delayed investigations and prosecutions on the effectiveness of justice systems is profound and far-reaching. It's not just a matter of inconvenience; it's a systemic failure that fuels violence, undermines the rule of law and erodes public confidence. The consequences go beyond individual cases, affecting entire communities and even destabilising nations. Delays allow criminals to continue their activities, often with impunity, emboldening them and fostering a climate of fear. This subsection examines the various facets of this problem, the factors that contribute to it and its devastating effects.

One of the most immediate consequences is the erosion of public confidence. When citizens see criminals remain at large for long periods despite evidence and arrests, trust in the justice system inevitably diminishes. This disillusionment breeds cynicism and can even lead to a rise in vigilantism and self-help justice, often leading to further violence and instability. In communities plagued by long delays, the feeling that the system is rigged against them - especially marginalised communities - often leads to frustration and anger, creating a fertile ground for social unrest.

The reasons for delays are many and complex. In many jurisdictions, resource constraints are a significant factor.

This requires robust mechanisms for judicial accountability, protection of judges and legal professionals from intimidation, and a commitment to transparency and the rule of law by all branches of government.

Understaffed police forces, overstretched prosecutors' offices and overburdened courts struggle to cope with the sheer volume of cases. This lack of resources often leads to investigative bottlenecks, where cases languish for lack of staff, equipment or expertise. In some countries, particularly those with limited technological resources, lack of access to forensic laboratories, digital forensics and advanced investigative tools significantly hampers the speed and effectiveness of investigations. This is particularly critical in cases involving complex crimes such as cybercrime, financial fraud or organised crime, where sophisticated investigative techniques are essential.

In addition, bureaucratic inefficiencies contribute significantly to delays. Complex procedures, cumbersome paperwork and a lack of inter-agency coordination can hold up investigations for months or even years. The movement of evidence, witness testimony and other critical information can be hampered by archaic systems and inadequate information technology. The resulting delays give criminals ample opportunity to hide evidence, intimidate witnesses and escape justice. The lack of proper case management systems often exacerbates the problem, with cases getting lost in the bureaucratic maze or simply forgotten amidst an overwhelming caseload.

Corruption plays a particularly insidious role in exacerbating delays. Bribery, influence peddling and other forms of misconduct can deliberately obstruct investigations and prosecutions. Judges can be bribed to delay decisions, prosecutors can be paid to drop charges, and police officers can be persuaded to turn a blind eye to criminal activity. In such an environment, justice is often not only delayed but denied altogether, creating a breeding ground for impunity and fostering a culture of lawlessness. This corrosive influence goes beyond individual acts of corruption and affects the very structure and legitimacy of the justice system. The prevalence of corruption erodes public confidence, demoralises honest officials and ultimately undermines the state's ability to maintain order and security.

The impact of delays on witnesses is of particular concern.

Witnesses may be reluctant to cooperate if they fear retaliation by criminals who are still active and at large. Memories fade, evidence is lost or destroyed, and the possibility of a successful prosecution diminishes with each passing day. Witness intimidation and harassment is particularly prevalent in cases involving organised crime, gang violence and political repression, further hampering the pursuit of justice. Witness protection programmes are often inadequate or non-existent in many jurisdictions, leaving witnesses vulnerable and exacerbating the problem of delayed justice.

The consequences go far beyond individual criminal cases. Delayed prosecutions undermine the deterrent effect of the criminal justice system. When criminals see that even serious crimes go unpunished for long periods of time, they are less likely to be deterred from future criminal activity.

This phenomenon leads to an escalation in crime rates, especially violent crime, as criminals feel emboldened by the apparent lack of consequences. The perception of impunity fosters a cycle of violence, with the lack of effective law enforcement and judicial accountability creating an environment in which criminal activity thrives.

The impact of delayed prosecution is particularly acute in cases involving vulnerable populations. Victims of domestic violence, sexual assault and human trafficking often face significant barriers to justice.

In seeking justice, delays only serve to exacerbate their suffering. These victims may already be traumatised and reluctant to testify, and the passage of time can further erode their willingness to participate in the legal process. Moreover, the delay itself can prolong their suffering and prevent them from moving on with their lives. In these cases, the failure of the justice system to act swiftly and decisively not only fails to protect victims, but can deepen the trauma they have endured.

Consider the case of a country grappling with a rising tide of gang violence. Years of delay in prosecuting gang members allow these groups to consolidate their power, expand their operations and recruit new members. The lack of effective law enforcement emboldens these gangs, leading to an increase in violent crime such as murder, extortion and drug dealing. The community is terrorised, businesses are intimidated and law-abiding citizens live in fear. The impunity enjoyed by these criminal organisations leads to a breakdown of social order and undermines efforts to promote sustainable economic development. The same dynamic can be observed in countries struggling with terrorism, organised crime and other forms of serious crime.

Addressing the problem of delays in investigations and prosecutions requires a multi-pronged approach. Increased investment in law enforcement and the judiciary is crucial, including adequate funding for personnel, training, technology and infrastructure. Streamlining bureaucratic procedures, improving inter-agency cooperation and implementing effective case management systems can significantly reduce delays. Robust anti-corruption measures are essential to ensure that the judiciary does not itself contribute to the problem. This includes establishing independent oversight bodies, strengthening accountability mechanisms and protecting whistleblowers who expose corrupt practices. Investing in witness protection programmes and providing support services for victims of crime is also crucial. Finally, building public confidence in the justice system through transparency, fairness and accessibility is paramount. By addressing the root causes of delays, strengthening the capacity of law enforcement and the judiciary, and building public trust, societies can work towards a more just and secure future. The effects of delayed justice reverberate through all strata of society, highlighting the critical importance of a speedy, efficient and impartial justice system for a peaceful and stable society. Failure to address these systemic failures leaves communities vulnerable to escalating violence and instability, underscoring the urgent need for comprehensive and sustained reform.

One of the most significant problems is the sheer length of time it takes for cases to reach trial. In many jurisdictions, particularly those struggling with overcrowded dockets and limited resources, the wait for a trial can be years or even decades. This long delay has several negative consequences. First, it allows offenders to remain at large and continue their criminal activities without fear of immediate consequences. This prolonged freedom reinforces a sense of impunity, which encourages them and may lead to further offences. The delay also allows criminals to intimidate witnesses, tamper with evidence and even flee the jurisdiction, making prosecution increasingly difficult or impossible.

The impact on victims is equally devastating. The long wait for justice prolongs their suffering and exacerbates the trauma they have endured. Victims may experience feelings of helplessness and frustration as their hopes for accountability and closure are repeatedly postponed. This delay can also have a negative impact on their mental and physical health, creating further challenges in their lives. The experience reinforces the sense that the justice system is failing to protect them, leading to a profound loss of faith in the rule of law.

Inadequate legal representation exacerbates the problems associated with trial delays. Many defendants, particularly those from marginalised communities or who lack financial resources, struggle to access competent legal counsel. This lack of access creates an uneven playing field where defendants without adequate representation are at a significant disadvantage in court.

They may be unable to effectively challenge evidence, present a robust defence or negotiate favourable plea bargains. This disparity undermines the fairness and impartiality of the judicial process and leads to wrongful convictions, further undermining public confidence in the system.

The inconsistency of sentencing practices between different jurisdictions, and even within the same jurisdiction, further contributes to a sense of impunity. Disparities in sentencing often reflect bias based on race, socio-economic status or other factors unrelated to the gravity of the crime committed. This arbitrariness in sentencing undermines the deterrent effect of the criminal justice system, as offenders are less likely to refrain from committing crimes if they perceive that the penalties are inconsistent or unfair.

The lack of clear and consistent sentencing guidelines, combined with the potential for judicial bias, creates an unpredictable and unfair system that fails to deliver proportionate punishment, leading to further escalation of criminal activity.

The consequences of these inefficiencies extend beyond individual cases. The erosion of public trust in the justice system is a significant societal cost. When citizens perceive the system as slow, inefficient or unfair, they are less likely to cooperate with law enforcement, report crimes or participate in the legal process. This loss of confidence in the system can lead to a breakdown in social order and an increase in self-help justice, potentially leading to further cycles of violence.

Consider the case of a country where trials routinely take several years and legal aid is severely limited. This system breeds resentment among citizens, particularly marginalised communities who experience disproportionate delays and lack of legal representation. As a result, individuals may feel increasingly justified in taking the law into their own hands, leading to vigilantism and potentially escalating conflicts. This cycle of mistrust and violence further strains the already fragile relationship between law enforcement and the public.

The problem is exacerbated by a lack of transparency and accountability within the justice system. In some jurisdictions, the lack of publicly available information on case processing times, court decisions and sentencing guidelines prevents meaningful oversight and accountability. This opacity contributes to public distrust and allows inefficiencies to persist.

Addressing these inefficiencies requires a multi-pronged approach. Investing in judicial infrastructure, increasing court staff and implementing streamlined case management systems can reduce delays in court proceedings. Expanding access to legal aid for indigent defendants is critical to ensuring a fair and impartial trial. Implementing standardised sentencing guidelines and providing ongoing training to judges on sentencing disparities can promote consistency and reduce bias. Finally, strengthening mechanisms for oversight and accountability within the justice system, including increased transparency and independent review of judicial decisions, will build public confidence and help ensure that the system effectively addresses violent crime and delivers justice fairly. Only by addressing these systemic shortcomings can we hope to break the cycle of violence and create a more just and equitable society.

The importance of addressing these issues cannot be overstated. Inefficient trials and inconsistent sentencing are not just technical problems; they are critical factors contributing to escalating violence and societal instability. They undermine the foundations of a just and functioning society by fostering a sense of impunity among criminals, undermining public confidence in the rule of law and perpetuating cycles of violence. A comprehensive reform agenda, including improvements in infrastructure, access to legal representation, sentencing guidelines and transparency, is needed to ensure that justice is not just delayed, but administered efficiently and fairly. Without such reforms, societies will continue to struggle with the devastating consequences of systemic failures in the justice process.

Compounding the problem is the lack of resources to effectively support victims throughout the justice process. Victims often face significant emotional, financial and logistical challenges during lengthy legal proceedings. Without adequate support services, such as counselling, financial assistance and victim advocacy, victims may be discouraged from actively participating in the legal process, thus hindering the prosecution's ability to secure a conviction. This lack of support reinforces the sense of injustice and deepens the trauma experienced by victims.

Investment in comprehensive victim support programmes is essential to ensure that justice is not only delivered, but also experienced as fair by those most affected by crime.

Furthermore, the interaction between inefficiencies in the judicial process and the overall political and social climate should not be overlooked. In countries experiencing political instability or widespread corruption, the justice system may be particularly vulnerable to manipulation and undue influence. This can lead to further delays, biased outcomes and a general erosion of public confidence in the ability of the legal system to deal effectively with violence. Strengthening the independence and integrity of the judiciary and insulating it from political pressure and corrupt influences is paramount to achieving a fair and effective system.

The erosion of public trust in institutions

A fundamental consequence, perhaps the most insidious, is the erosion of public confidence in institutions, particularly the judiciary and, by extension, government as a whole. This loss of confidence undermines the very foundations of a stable and peaceful society, creating a fertile ground for further instability and violence. The perception - often justified - that the justice system is broken, biased or simply ineffective breeds cynicism and despair, leaving citizens feeling abandoned and powerless.

This erosion of trust is not just a matter of public opinion; it has tangible consequences. When citizens lose faith in the ability of the legal system to protect them, redress grievances and punish wrongdoers, they are less likely to cooperate with law enforcement, report crimes or participate in community initiatives to promote safety and security. This creates a vicious cycle: a lack of trust leads to less reporting, which in turn allows criminal activity to flourish, further undermining confidence in the effectiveness of the system.

The problem is compounded by the inherent inequalities that are often embedded in justice systems. In many societies, access to justice is disproportionately distributed, favouring the wealthy and powerful while leaving marginalised groups vulnerable to exploitation and abuse. When those with resources are able to navigate the system more effectively and secure speedy justice and favourable outcomes, while those without resources face long delays and inadequate legal representation, perceptions of injustice are reinforced. This disparity fuels resentment and mistrust, particularly among those who already feel marginalised and disenfranchised.

This perception of bias, whether real or perceived, is significantly exacerbated by media coverage. Sensationalised reporting of high-profile cases where justice is delayed or appears to favour certain groups can significantly damage public trust. The 24/7 news cycle, amplified by social media, ensures the rapid dissemination of such narratives, potentially shaping public opinion in a way that reinforces negative perceptions of the justice system. This effect is particularly strong when the media focus on cases of police misconduct or allegations of corruption within the judiciary itself, further undermining confidence in the integrity of the institutions charged with upholding the law.

When justice is denied to victims of human rights abuses, the resulting resentment and anger can fuel cycles of violence for generations. Similarly, in countries with weak or corrupt judicial systems, the lack of effective law enforcement and the perception of impunity for powerful elites often leads to widespread social unrest and violent protests.

The erosion of trust is not limited to developing countries. Even in established democracies, lengthy delays in judicial processes, coupled with perceptions of bias or corruption, can undermine public confidence. This can manifest itself in declining civic participation, increased political polarisation and a growing sense of alienation from state institutions. The recent rise of populist movements in many parts of the world can be partly attributed to widespread dissatisfaction and disillusionment with established political and judicial systems. Citizens who feel unheard and ignored, who perceive a lack of fairness and accountability, are more susceptible to the rhetoric of populist leaders who promise radical change and a return to traditional values.

Restoring public confidence requires a multi-pronged approach that addresses both the structural deficiencies within the justice system and the underlying social and political factors that contribute to disillusionment. First, significant reforms are needed to address systemic problems such as court backlogs, inadequate funding and lack of access to legal representation. Improving transparency and accountability within the judiciary is crucial, including measures such as independent oversight bodies and mechanisms for effective complaints handling. Second, measures to promote diversity and inclusiveness within the judiciary are essential. This includes ensuring representation of marginalised groups at all levels of the judiciary and law enforcement, and developing policies to address systemic bias.

It is also essential to improve communication and engagement between the justice system and the public. This includes initiatives aimed at improving public understanding of the justice system, promoting greater transparency in judicial processes, and establishing mechanisms for direct feedback and engagement. It is also essential to promote a culture of respect for the rule of law and human rights. This requires investment in education and civic engagement programmes, promoting greater awareness of legal rights and responsibilities, and encouraging participation in community-based initiatives to promote social cohesion.

It is also essential to address the socio-economic factors that underpin mistrust. Reducing inequality, promoting economic opportunity and strengthening social safety nets can go a long way towards reducing public discontent and fostering a greater sense of fairness and justice. In addition, addressing underlying grievances and actively engaging with communities affected by past injustices can help rebuild trust and foster a sense of reconciliation. This requires acknowledging past wrongs, providing effective reparations to victims, and implementing policies that promote reconciliation and social healing.

Finally, the role of the media in shaping public perceptions cannot be overlooked. While media scrutiny is essential to ensure accountability, responsible reporting is essential to prevent the spread of misinformation and the reinforcement of negative stereotypes. Promoting accurate and balanced reporting, while highlighting positive developments and examples of effective justice, can help restore public confidence in the system. Journalists need to play a constructive role, focusing on investigative reporting that exposes corruption and bias, while also highlighting stories of successful prosecutions and effective remedies.

The erosion of public trust in institutions is a serious threat to social stability and security. It is a complex problem with deep roots that requires a multi-faceted and sustained response. Addressing shortcomings in the justice system, promoting inclusiveness and fairness, improving communication and engagement, and tackling underlying socio-economic inequalities are critical steps in rebuilding trust and creating a more just and peaceful society. Failure to do so will only perpetuate the cycle of violence and instability, undermining the foundations for a stable and prosperous future.

Improving investigative capacity and effectiveness

Improving investigative capacity and efficiency is critical to reducing violence and promoting a climate of justice. Inefficient investigations contribute significantly to a sense of impunity, emboldening criminals and undermining public confidence. Delays in collecting evidence, processing crime scenes and arresting suspects allow perpetrators to evade arrest and potentially commit further crimes. This prolonged period of uncertainty also causes significant distress to victims and witnesses and may deter them from cooperating with the authorities. However, a robust and efficient investigation system acts as a powerful deterrent, signalling that criminal acts will not go unpunished. It also contributes to a sense of security and public order, reducing the likelihood of further violence.

One of the key aspects of improving investigative capacity is a significant increase in the resources allocated to law enforcement agencies. This includes not only funding for equipment and technology, but also investment in human capital. Many jurisdictions struggle with understaffing, overworked detectives and a lack of specialised units to deal with specific types of crime. This leads to a backlog of cases, delays in investigations and reduced effectiveness in solving crimes. Adequate funding should allow for increased recruitment, better pay and benefits to attract and retain qualified staff, and the establishment of specialist units focusing on areas such as cybercrime, domestic violence or organised crime. In addition, this increased investment should prioritise providing investigators with the necessary tools and technology to conduct effective investigations. This includes advanced forensic equipment, sophisticated surveillance technology and access to national and international databases. Modern communication systems and software can also significantly streamline investigative processes, allowing for better collaboration and data management.

Improving the training of law enforcement personnel is just as important as increasing resources. Many investigations fail not because of a lack of resources, but because of inadequate training and professional development. Investigators need thorough training in forensic science, interviewing techniques, evidence collection and preservation, and legal procedures. This training should be continuous and regularly updated to reflect changes in criminal tactics and technological advances. Specialised training should be provided for specific types of crime to enable investigators to better understand the nuances of different criminal activities. Regular refresher courses and training opportunities are essential to maintain the skills and knowledge of experienced officers. Similarly, training should emphasise ethical behaviour and integrity, as a culture of professionalism is essential to ensure the legitimacy and credibility of investigations. This includes training on how to identify and mitigate bias in investigations to ensure fair and impartial outcomes.

Beyond individual training, emphasis should be placed on improving coordination and cooperation between different law enforcement agencies and other relevant stakeholders. Many jurisdictions struggle with fragmented law enforcement structures, resulting in a lack of information sharing and effective communication. This can significantly hamper investigations, particularly in cases involving organised crime or cross-border activities. Improved communication protocols, joint investigation teams and the creation of centralised databases are essential to promote effective cooperation. This inter-agency cooperation must go beyond law enforcement and include other relevant agencies such as prosecutors, forensic laboratories and victim support services. Regular meetings, joint training and the development of standardised procedures are essential to achieve seamless cooperation. Equally important is the establishment of clear lines of authority and responsibility to ensure accountability and transparency in the investigative process.

Examining successful reforms in different jurisdictions provides valuable insights into effective strategies for improving investigative capacity. For example, the United Kingdom has made significant investments in forensic science and technology that have led to significant improvements in crime detection and conviction rates. Similarly, countries such as Singapore have implemented sophisticated crime analysis and prediction systems, enabling them to proactively target high-risk areas and prevent crime. These successes demonstrate the tangible impact of investment in technology, training and inter-agency coordination. These successful models provide valuable templates for other jurisdictions to adapt and implement, tailoring their approach to their specific circumstances and resources. These case studies offer valuable lessons, highlighting the potential for significant improvements in investigative capacity and effectiveness through strategic allocation of resources and systematic improvements in training and coordination. The measurable improvements in crime reduction and increased conviction rates provide compelling evidence of the positive impact of these reforms.

However, it is important to recognise that improving investigative capacity is not a stand-alone solution to reducing violence. It must be part of a broader strategy that addresses the underlying social and economic factors that contribute to crime. Reducing poverty, improving education and access to health care, and promoting employment opportunities are critical to addressing the root causes of violence. Without addressing these issues, even the most effective criminal justice systems may not be able to prevent the recurrence of violent crime. Therefore, a comprehensive approach that combines improvements in investigative capacity with broader social and economic development initiatives is essential for sustainable reductions in violence. This holistic strategy requires a coordinated effort involving various government agencies, non-profit organisations and international partners to address the interrelated challenges of reducing violence and improving the overall well-being of the community.

The focus should also extend to ensuring accountability within law enforcement agencies. Cases of police misconduct and corruption significantly undermine public confidence and the effectiveness of investigations. Independent oversight bodies with the power to investigate allegations of misconduct and prosecute offenders are essential to promote transparency and accountability.

These bodies should be impartial and independent of law enforcement agencies to ensure objective and thorough investigations. In addition, the establishment of robust internal affairs systems within police departments can help prevent misconduct and address any problems promptly and effectively. Regular audits, transparent reporting mechanisms and clear ethical guidelines are essential to maintain high professional standards within law enforcement.

Strengthening prosecution and ensuring fair trials

Strengthening law enforcement is critical to reducing violence. Weak prosecution, characterised by insufficient resources, inadequate training and a lack of prosecutorial independence, contributes significantly to a climate of impunity.

This allows criminals to operate with little fear of legal consequences, fostering a culture of violence and undermining public confidence in the justice system. Conversely, strong and effective prosecutions act as a powerful deterrent, signalling that criminal acts will not go unpunished. This deterrent effect is crucial in reducing crime rates and promoting a sense of safety and order in the community.

A key aspect of strengthening law enforcement is to provide prosecutors with sufficient resources. This includes adequate funding for staff, training, technology and infrastructure. Understaffed prosecutor's offices, often overwhelmed by large caseloads, are unable to effectively investigate and prosecute crimes. This leads to plea bargains to avoid trial, even in serious cases, resulting in reduced sentences and inadequate punishment. Increased funding should allow for the recruitment of more prosecutors, investigators and support staff, thereby reducing caseloads and allowing for more thorough investigations and prosecutions.

It should also be used to acquire modern technology, such as case management systems and forensic analysis software, to improve the efficiency and effectiveness of the prosecution process.

In addition to financial resources, investment in training is equally important. Prosecutors need specialised training in various areas, including legal procedure, presentation of evidence, courtroom tactics and negotiation skills. This training needs to be continuous and regularly updated to reflect changes in the law and in criminal tactics.

Specialised training is needed to deal with complex cases, such as organised crime, cybercrime or domestic violence, which require specific expertise and knowledge. In addition, training should emphasise ethical behaviour and respect for the integrity of the judicial process. This will ensure that prosecutors uphold the highest standards of professionalism and fairness and protect the rights of both victims and defendants.

Ensuring prosecutorial independence is another essential component of a strong and effective prosecution system. Prosecutors must be free from political interference and undue influence from other branches of government or powerful individuals. This independence is essential to ensure that cases are prosecuted on the merits of the evidence and the law, and not on the basis of political considerations or personal agendas. Measures to ensure prosecutorial independence include establishing clear guidelines for prosecutorial conduct, providing adequate protection for prosecutors from intimidation or retaliation, and empowering oversight bodies to investigate allegations of political interference or corruption.

Fair and impartial trials are the cornerstone of any fair and effective justice system. However, many judicial systems suffer from delays, procedural irregularities and lack of access to competent legal representation, which undermine the fairness and effectiveness of the trial process. These shortcomings often result in wrongful acquittals or lenient sentences, which encourage further criminal activity and fuel public dissatisfaction with the justice system. Addressing these problems requires a multi-pronged approach, including reforms to trial procedures, increased access to legal aid and improved training for the judiciary.

Reducing delays in court proceedings is crucial to ensuring a fair and efficient judicial process. Lengthy delays contribute to a sense of impunity, allowing perpetrators to remain at large and potentially commit further crimes. Delays also cause distress to victims and witnesses and may discourage them from cooperating with the justice system.

In addition, long delays can lead to the deterioration of evidence, making it more difficult to secure convictions. Efficient case management, improved court administration and technological advances, such as electronic filing systems, can significantly reduce delays in court proceedings. Prioritising cases according to their seriousness and using specialised courts to deal with certain types of crimes can also improve efficiency.

Access to competent legal representation is a fundamental right in any just society. However, many defendants, particularly those from disadvantaged backgrounds, do not have access to adequate legal representation.

This inequality undermines the fairness of the judicial process and leads to disproportionate convictions and unjust sentences. Increasing funding for legal aid programmes, providing training and support for public defenders, and ensuring access to pro bono services from private lawyers are critical to addressing this issue.

In addition, establishing clear standards for legal representation to ensure that all defendants are adequately represented is critical to ensuring a fair trial.

Improving the quality of judicial training is also essential. Judges must have a thorough understanding of the law, procedure and evidence, as well as the skills to conduct fair and impartial trials. Regular judicial training programmes covering substantive law, procedure, ethics and judicial temperament can greatly improve the quality of justice. In addition, training should include sensitivity training to address issues of bias and impartiality and to ensure that judges make decisions based solely on the evidence presented. These improvements in judicial training can significantly enhance the fairness and efficiency of the judicial process.

Analysis of successful judicial reforms in other jurisdictions can provide valuable lessons and insights into effective strategies. For example, countries that have established specialised courts for particular types of crime, such as domestic violence or drug trafficking, have often seen significant improvements in prosecution rates and reduced delays. Similarly, countries that have invested in advanced technology and case management systems have seen improvements in the efficiency of their justice systems. It is also crucial to recognise the importance of community engagement and participation in promoting justice and reducing violence. Restorative justice programmes, community mediation and victim support services can play an important role in addressing the root causes of crime and promoting reconciliation within communities. These programmes aim to repair the harm caused by crime and promote a sense of responsibility and accountability among offenders. By working with the community, the justice system can better understand the specific needs and challenges faced by different groups and tailor its approach accordingly. This collaborative approach can lead to more effective crime prevention strategies and a fairer and more just society.

Finally, the effectiveness of justice reforms must be rigorously monitored and evaluated. This involves collecting data on key indicators such as prosecution rates, trial delays, conviction rates and recidivism rates in order to assess the impact of the reforms implemented.

Regular evaluations provide valuable feedback that enables adjustments and improvements to be made to the strategies implemented, maximising their effectiveness and ensuring that they achieve their intended goals. Ongoing monitoring and evaluation ensures that the justice system remains adaptable and responsive to evolving needs and challenges, ultimately contributing to the reduction of violence and the promotion of a more just and peaceful society.

Transparent and accessible reporting of these evaluations is also crucial to fostering public confidence in the justice system.

Policy recommendations for reducing violence

A multi-pronged approach is needed, including legislative reform, institutional strengthening and societal changes in attitudes towards justice. One of the most pressing issues to be addressed is the pervasive problem of judicial backlogs. The sheer volume of unresolved cases in many judicial systems around the world creates a climate of impunity in which perpetrators of violence, especially those with resources, can escape justice for long periods of time. This delay undermines the deterrent effect of the law and encourages further criminal activity. A key policy recommendation is therefore to invest heavily in increasing judicial capacity. This includes not only increasing the number of judges and court personnel, but also improving their training, technology and infrastructure. Efficient case management systems, including streamlined procedures, electronic filing and robust data analysis, are essential to reduce delays and ensure that cases are dealt with expeditiously. This is particularly important in cases of violence, where victims need timely justice to heal and move on. International organisations and donor agencies should prioritise funding initiatives focused on improving court efficiency and technological modernisation in countries with significant case backlogs.

Strengthening police investigative capacity is also essential.

Many justice systems fail because of weak investigations, leading to insufficient evidence and the dismissal of cases. Investment in forensic science, specialised investigative units and training programmes for law enforcement personnel can significantly improve the quality of investigations and the likelihood of successful prosecutions. In addition, improved cooperation between police and prosecutors is essential to avoid fragmented and ineffective investigations that impede justice. This inter-agency coordination should be supported by joint training, shared data systems and clear protocols for information sharing. This cooperation is particularly important in the prosecution of organised crime and complex cases involving violence, as it allows for a coordinated and comprehensive approach to evidence gathering and prosecution.

Tackling corruption within the justice system is another critical element. Corruption undermines public confidence and allows powerful individuals and criminal networks to go unpunished. Strengthening the independence of the judiciary, implementing robust anti-corruption mechanisms and establishing transparent and accountable systems for judicial appointments are essential steps. This includes increasing the salaries and benefits of judicial officials to reduce their vulnerability to bribery, strengthening internal disciplinary mechanisms for judicial misconduct, and empowering independent oversight bodies to investigate allegations of corruption.

In addition, the protection of whistleblowers who expose corruption within the judiciary is essential to ensure accountability and prevent the erosion of public confidence. International organisations can play an important role by supporting the establishment of independent anti-corruption commissions and providing technical assistance in the development of effective anti-corruption strategies.

The issue of bias and discrimination within the judicial system also requires attention. Implicit and explicit bias can lead to unfair and inconsistent outcomes, particularly for marginalised communities and victims of gender-based violence. Implementing measures to combat bias requires a multi-faceted approach. This includes increasing diversity within the judiciary, promoting gender equality and providing comprehensive training on unconscious bias to judges, prosecutors and other court personnel. In addition, the use of data analysis and monitoring mechanisms can help to identify and address disparities in judicial outcomes. Regular audits of judicial decisions, combined with the use of statistical tools, can reveal patterns of bias and inform targeted interventions. This data-driven approach can help ensure fairness and equity within the justice system and prevent the perpetuation of systemic discrimination.

Improving access to justice for victims of violence is another critical aspect of reducing violence in society. Many victims, particularly those from marginalised communities, face significant barriers to accessing legal assistance, including financial constraints, geographical limitations and cultural barriers. Policy recommendations include strengthening legal aid programmes, providing free or subsidised legal services to victims of violence, and establishing easily accessible and user-friendly legal services. This could include expanding the reach of pro bono legal services, employing legal professionals who understand the specific needs of marginalised communities, and using technological tools to overcome geographical barriers. In addition, the provision of support services for victims, including counselling, medical care and witness protection programmes, is essential to improve their experience of the justice system and reduce barriers to seeking justice.

Promoting a culture of accountability is crucial to reducing violence. This requires not only holding perpetrators of violence accountable through effective prosecution, but also holding justice officials accountable for their performance and integrity. The implementation of effective monitoring mechanisms, including performance and disciplinary procedures, is crucial. Public oversight and feedback mechanisms should also be put in place to allow citizens to voice their concerns and report instances of misconduct or inefficiency. Transparency in judicial proceedings and decisions is also essential to build public confidence. This can be achieved by publishing judicial decisions, using open data initiatives and providing access to information on the functioning of the judicial system. Strengthening the role of the media in holding the judiciary accountable is also crucial. Freedom of the press allows for scrutiny of the justice system, highlighting cases of corruption, bias or inefficiency.

Finally, it is essential to invest in comprehensive crime prevention strategies. While effective justice systems are crucial for dealing with violence once it has occurred, a comprehensive approach also requires preventive measures. These include investment in education, poverty reduction programmes, community development initiatives and initiatives to promote conflict resolution and social cohesion. Addressing the underlying social and economic factors that contribute to violence is crucial to reducing crime rates in the long term. This requires cross-sectoral cooperation, bringing together government agencies, community organisations and international partners to implement comprehensive crime prevention strategies tailored to the specific needs of each community. This multifaceted approach, combined with the strengthening of the justice system, can effectively address societal violence.

Bibligraphy

Adler, L.L. & Denmark, F.L. (Eds.). (1995). Violence and the Prevention of Violence. Westport, Ct: Praeger Publishers.

Apter, D.E. (Ed.). (1996). The Legitimization of Violence. New York: Washington: New York University Press.

American Psychological Association. "How Can We Intervene Effectively? In the Report from the Commission on Violence and Youth. Violence and Youth: Psychology's Response, New York: Sol Charitable Trust. (pp. 53-62)

Apfel, R.J. & Simon, B. (Eds.), (1996). Minefields in their Hearts: the Mental Health of Children in War and Communal Violence. New Haven: Yale University Press.

Bandarage, A. (1997). Women, Population and Global Crisis: A Political-Economic Analysis. London: Zed Books.

Blair, J. (1994). Making Behavioral Change the outcome Measure in Research on Violence. Psychological Reports, 75 (3, pt.1)

Busby, D.M. (1996). The Impact of Violence on the Family: Treatment Approaches for Therapists and Other Professionals. Boston: Allyn & Bacon.

Cahn, D.D. & Lloyd, S.A. (Eds.), (1996). Family Violence from a Communication Perspective. Thousand Oaks, CA: Sage.

Cairns, Ed. (1996). Children and Political Violence. Cambridge, Mass: Blackwell Publishers.

Cameron, D. & Frazer, E. (1987). The Lust to Kill: A Feminist Investigation of Sexual Murder. New York: University Press.

Carr, S.C. & Schumaker, J.F. (1996). Psychology and the Developing World. Westport: Praeger.

Cleaver, R. & Myers, P. (Eds.), (1993). A Certain Terror: Heterosexism, Militarism, Violence & Change. Chicago: Great Lakes Regional Office, American Friends Service Committee, 1993.

Cohen, L. & Swift, S. (1993). A Public Health Approach to the Violence Epidemic in the United States. Environment and Urbanization, 5 (2) 50-66.

Corrin, C. (Ed.). (1996). Women in a Violent World: Feminist Analyses and Resistance Across 'Europe'. Edinburgh: Edinburgh University Press.

Davey, J. D. (1995). The New Social Contract: America's Journey from Welfare State to Police State. Westport, CT: Praeger Publications.

Davies, M., (Ed), (1994). Women and Violence: Realities and Responses Worldwide. Zed Books.

De Zulueta, F. (1993). From Pain to Violence: The Traumatic Roots of Destructiveness. London: Whurr, Publishers, Ltd.

De Levita, D. (1995). Treatment of War-Traumatised Children in Former-Yugoslavia. In Groen-Prakken, H. & Ladan, A. & Stufkens, A. (Eds.), The Dutch Annual of Psychoanalysis 1995-1996: Traumatisation and War, Vol. 2 (pp.165-180). Swets & Zeitlinger: Amsterdam.

Desjarlais, R. et al. (1995). World mental health: problems, and priorities in low-income countries. New York : Oxford University Press.

Doner, K. (1994). Why Violence is a Health Care Priority. Social Policy, 24 (3) 58-62.

Erlich, H.J. (1990). The Ecology of Anti-Gay Violence. Special Issue: Violence Against Lesbians and Gay Men: Issues for Research and Practice. Journal of Interpersonal Violence, 5(3) 359-365.

Eron, L.D., Gentry, J.H., & Schlegel, P. (1994). Reason to Hope: A Psychosocial Perspective on Violence & Youth. Washington, DC: American Psychological Association.

Fawcett, B., Featherstone, J.H. & Toft, C. (1996). Violence and Gender Relations: Theories and Interventions. Thousand Oaks, CA: Sage.

Fiorenza, E.S. & Copeland, S. (1994). Violence Against Women. Mary Knoll, New York: Orbis Books.

Freedy, J.R. & Hobfoll, S.E. (1995). Traumatic Stress: From Theory to Practice. New York: Plenum Press.

Foucault, M., "Two Lectures", In: Colin Gordon (ed.), Power/Knowledge – Selected Interviews and Other Writings 1972-1977, Pantheon Books, New York: 1980 (1976a).

Foucault, M., The Will to Knowledge - The History of Sexuality vol. 1, Penguin Books, London: 1998 (1976b).

Foucault, M., "On the Genealogy of Ethics – an Overview of a Work in Progress", In: Rabinow, Paul (ed.), The Foucault Reader, pp. 340-372, Pantheon Books, New York: 1984 (1983).

Foucault, M., "The Ethic of Care for the Self as a Practice of Freedom", In: Fornet-Betancourt, R., Becker, H. and Gomez-Müller, A. (conduction), Philosophy Social Criticism, 12(2-3), pp. 112-131, Sage Publications: July, 1987 (1984).

Galtung, J. "Violence, Peace and Peace Research", In: Journal of Peace Research, Vol. 23, No. 9, Sage Publications: 1968.

Galtung, J., "Cultural Violence", In: Journal of Peace Research, Vol. 27, No. 3, Sage Publications: 1990.

Grossman, D. (1995). On Killing: The Psychological Cost of Learning to Kill In War and Society. Boston: Little, Brown.

Griffiths, M.D. & Shuckford, G.L. (1989). Desensitization to Television Violence: A New Model. New Ideas in Psychology, 7(1), 85-89.

Hunt, W.B. (1997). Getting to War: Predicting International Conflict with Mass Media Indicators. Ann Arbor, MI: University of Michigan Press.

James, C.B. (1993). Race, Class, Violence and Health in an Urban Ghetto. Manchester, England:Nuevan.

Harvery, J.H. (1996). Embracing their Memory: Loss and the Social Psychology of Storytelling. Boston, MA: Allyn & Bacon, Inc..

Hamm, M.S. & Schrink, J.L. (1989). The Conditions of Effective Implementation: A Guide to Accomplishing Rehabilitative Objectives in Corrections. Criminal Justice and Behavior, 16 (2) 166-182.

Hammond, W.R. & Yung, B. (1993). Psychology's Role in the Public health Response to Assaultive Violence among Young African-American Men. American Psychologist, 48, 151-154.

Heise, L. (1994). Violence Against Women: the Hidden Health Burden. Washington, DC: World Bank.

Heise, L. (1995). Violence, Sexuality and Women's Lives. In Parker, R. G. & Gagnon, J.H. (Eds.), Conceiving Sexuality: Approaches to Sex Research in a Postmodern World. (pp.109-134). New York: Routledge.

Hickson, J. & Kriegler,S. (Eds.), (1996). Multicultural Counseling in a Divided and Traumatized Society: the Meaning of Childhood and Adolescence in South Africa. Westport, Conn.: Greenwood Press.

Keane, T.M., Albano, A.& Blake, D.D. (1992). Current Trends in the Treatment of Post-Traumatic Stress Symptoms. In Basoglu, M. (Ed.), Torture and its Consequences: Current Treatment Approaches (pp.363-401). Cambridge, England: Cambridge University Press.

Kime, Z.R. (1992). Aberrant Sexual Behavior, Violence, and Reproduction. Journal of Sex and Education Therapy, 18 (4) 231-241.

Kleber, R.J., Figley, C. R., & Gersons, B. P. R., (Eds.), (1995). Beyond Trauma: Cultural and Societal Dynamics. New York, NY: Plenum Press.

Kleck, G. (1991). Point Blank: Guns and Violence in America. New York: A. De Gruyter.

Lykes, M.B., Brabeck, M.M., Ferns, T. & Radan, A. (1993). Human Rights and Mental Health Among Latin American Women in Situations of State Sponsored Violence: Bibliographic Resources. Psychology of Women Quarterly, 17(4), 525-544.

Nirenberg, D. (1996). Communities of Violence: Persecution of Minorities in the Middle Ages. Princeton, NJ: Princeton University Press, 1996.

"Preventing Violence and Abusive Behavior: A Public Health Agenda" Proceedings of the Seventh National Conference on Health Care for the Poor and Underserved at Meharry Medical College Nashville, Tennessee. In Conjunction with The Lloyd C. Elam Mental Health Symposium and " I Have a Future" Adolescent Health Promotion Program. In Journal of Health Care for the Poor and Underserved, 6(2) 1-300.

Prothrow-Stith, D. (1990). The Epidemic of Violence and its Impact on the Health Care System. Henry Ford Hosp Med J., 38 (2&3), 175-177.

Prothrow-Stith, D. (1991). Deadly Consequences: How Violence Is Destroying our Teenage Population and a Plan to Begin Solving the Problem. New York, NY: Harper Collins.

Prothrow-Stith, D. (1992). Can Physicians Help Curb Adolescent Violence? Hospital Practice, 6, 193-207.

Prothrow-Stith, D.& Spivak, H. (1990). Violence. In McAnarney, E. (Ed.), Adolescent Medicine (pp.1113-1118). Philadelphia: W.B. Saunders.

Rapoport, A.(1989). The Origins of Violence: Approaches to the Study of Conflict. New Brunswick, N.J: Transaction Publishers.

Reiber, R.W. (1997). Manufacturing Social Distress: Psychopathy in Everyday Life. New York: Plenum Press.

Reiss, A.J. & Roth, J.A. (1993-1994). Understanding and Preventing Violence. Washington, DC: National Academy Press. (4 volumes. v.2 Biobehavioral Influences, v.3 Social Influences, v.4 Consequences and control.)

Rosenberg, M.L. & Fenley, M.A. (1991) Violence in America: A Public Health Approach. New York: Oxford University Press.

Ruback, R.B. & Weiner, N.A. (Eds.) (1995). Interpersonal Violent behaviors: Social and Cultural Aspects. New York: Springer Pub. Co..

Saper, B. (1988). On Learning Terrorism. Terroroism, 11(1), 13-27.

Shalala, D. (1993). Addressing the Crisis of Violence. Health Affairs, 12 (4), 30-33.

Saunders, D.G. (1984). Helping Husbands Who Batter. Social Casework, 65 (6), 347-353.

Scherper-Hughes, N. (1992). Death Without Weeping: the Violence of Everday Life in Brazil. Berkeley. University of California Press.

Segal, H. (1997). Psychoanalysis, Literature and War: Papers 1972-1995. New York: Routledge.

Shrader, W.K. (1992). Media Blight and the Dehumanizing of America. New York: Praeger Publishers.

Sidanius, J. & Liu, J.H. (1992). The Gulf War and the Rodney King Beating: Implications of the General Conservatism and Social Dominance Perspectives. Journal of Social Psychology, 132 (6) 685-700.

Sidel, R. (1992). Women and Children Last: the Plight of Poor Women In Affluent America. New York: Penguin Books.

Starrels, M.E. (1994). The Feminization of Poverty in the United States: Gender, Race, Ethnicity and Family Factors. Journal of Family Issues. 15, 605-607.

Sugarman, A. (1994). Victims of Abuse: the Emotional Impact of Child and Adult Truama. Madison, CT: International Universities Press.

Turpin, J. & Kurtz, L.R., (Ed.), (1997). The Web of Violence: From Interpersonal to Global. Chicago, Ill.: University of Illinois Press.

Ulman, R.B. (1988). The Shattered Self: A Psychoanalytic Study of Trauma. Hillsdale, NJ: Analytic Press.

Walsh, D.A. (1994). Selling Out America's Children: How America Puts Profit Before Values and What Parents Can Do. Minneapolis, Minn.: Deaconess Press.

Ward, C. (1995). Attitudes Toward Rape: Feminist and Social Psychological Perspectives. London: Sage Publications.

Whalen, M. (1996). Counseling to End Violence Against Women: A Subversive Model. Thousand Oaks, CA. Sage Publications.

Whitebook, Joel. (1995). Perversion and Utopia: A Study in Psychoanalyis and Critical Theory. Cambridge, MA: MIT Press.

Widom, C.S. (1989). "Does Violence Beget Violence?" In Weiner, N.A. & Wolfgang, M.E. (Eds.), Pathways to Criminal Violence. Newbury Park, CA: Sage Publications.

Young, P. & Jesser, P. (1997). The Media and the Military From the Crimea to Desert Strike. New York: St. Martin's Press.

About the Author

Artur Vitória is a Portuguese writer and involved in several associative activities

Education – He graduated in Law from the Faculty of Law of the University of Lisbon in 1974 and is a graduate of the National Defence Institute in Lisbon, Portugal (1990). Dedicated to his academic career, in 1980 he became a researcher at the Foreign Relations Observatory of the Autonomous University of Lisbon.

Commends:

D. Afonso Henriques Medal - Army Merit 2nd Class with Commendation on 6th November 2015.

Medal of the Cross of Saint George on 18 June 2020.

honorary orders

Knight of the Order of Rio Branco on 7 December 2022.

Knight of the Order of Naval Merit on 11 June 2019.

Officer of the Order of Naval Merit on 11 June 2021.

Knight of the Order of Aeronautical Merit on 23 October 2016.

Officer of the Order of Aeronautical Merit on 23 October 2020.

Read more at https://observare.autonoma.pt/investigador/artur-victoria/.

www.ingramcontent.com/pod-product-compliance
Lightning Source LLC
LaVergne TN
LVHW082336200926
848481LV00012B/3360